How to Become a Train Driver

www.How2Become.com

As part of this product you have also received FREE access to online tests that will help you to pass the Train Driver Tests.

To gain access, simply go to:

www.MyPsychometricTests.co.uk

Get more products for passing any test at:

www.How2Become.com

Orders: Please contact How2Become Ltd, Suite 1, 60 Churchill Square Business Centre, Kings Hill, Kent ME19 4YU.

You can order through Amazon.co.uk under ISBN: 9781909229501, via the website www.How2Become.com or through Gardners.com.

ISBN: 9781909229501

First published in 2015 by How2Become Ltd.

Updated in 2018 by How2Become Ltd.

Typeset for How2Become Ltd by Gemma Butler.
Printed and bound by CPI Group (UK) Ltd,
Croydon, CR0 4YY

Disclaimer

Every effort has been made to ensure that the information contained within this guide is accurate at the time of publication. How2Become Ltd is not responsible for anyone failing any part of any selection process as a result of the information contained within this guide. How2Become Ltd and their authors cannot accept any responsibility for any errors or omissions within this guide, however caused. No responsibility for loss or damage occasioned by any person acting, or refraining from action, as a result of the material in this publication can be accepted by How2Become Ltd.

The information within this guide does not represent the views of any third party service or organisation.

CONTENTS

INTRODUCTION

Dear Candidate,

Welcome to *How to Become a Train Driver: The Insider's Guide*. This guide has been designed to help you prepare for and pass the Trainee Train Driver selection process. We feel certain that you will find the guide both a comprehensive and highly informative tool for helping you obtain one of the most sought after careers available.

The Train Driver selection process is not easy. It is comprehensive, relatively drawn out and highly competitive. In fact, on average there are between 300 and 400 applicants for every vacancy. Coupled with the fact that Train Operating Companies rarely advertise posts, this makes it an even harder job to obtain. However, do not let this put you off as many of the applicants who do apply are grossly under prepared, and they normally fail at the first hurdle. You must prepare fully if you are to pass the selection process and be offered a position as a Trainee Train Driver. There are a number of things that you can do in order to increase your chances of success, and they are all contained within this guide. The majority of Train Operating Companies (TOCs) are both professional and meticulous in how they run their assessment centres, and you should find the process an enjoyable one. We hope that you enjoy the guide and we wish you all the best in your pursuit of becoming a Train Driver.

If you would like any further assistance with the selection process then we offer the following products and training courses via the website www.How2Become.com:

- How to complete the Train Driver application form DVD;

- How to pass the Train Driver interview DVD;

- 1-Day intensive Train Driver course.

Finally, you won't achieve much in life without hard work, determination and perseverance. Work hard, stay focused and be what you want!

Good luck and best wishes,

The how2become team

The How2Become team

- -

Attend a 1-Day
Train Driver Course at:
www.TrainDriverCourse.co.uk

- -

PREFACE BY AUTHOR RICHARD McMUNN

Before I get into the guide and teach you how to prepare for the Trainee Train Driver selection process, it is important that I explain a little bit about my background and why I am qualified to help you succeed.

I joined the Royal Navy soon after leaving school and spent four fabulous years in the Fleet Air Arm branch onboard HMS Invincible. It had always been my dream to become a Firefighter and I only ever intended on staying in the Royal Navy for the minimum amount of time. At the age of 21 I left the Royal Navy and joined Kent Fire and Rescue Service. Over the next 17 years I had an amazing career with a fantastic organisation. During that time I was heavily involved in training and recruitment, often sitting on interview panels and marking application forms for those people who wanted to become Firefighters. I also worked very hard and rose to the rank of Station Manager. I passed numerous assessment centres during my time in the job and I estimate that I was successful at over 95% of interviews I attended.

The reason for my success was not because I am special in any way, or that I have lots of educational qualifications, because I don't! In the build up to every job application or promotion I always prepared myself thoroughly. I found a formula that worked and that is what I intend to teach you throughout the duration of this book.

Over the past few years I have taught many people how to pass the selection process for becoming a Trainee Train Driver, both through this guide and also during my one day intensive training course at www.traindrivercourse.co.uk. Each and every one of the students who attends my course is determined to pass, and that is what you will need to do too if you are to be successful. As you are probably aware many people want to become Train Drivers. As a result of this, the competition is fierce. However, the vast majority of people who do apply will submit poor application forms or they will do very little to prepare for the assessment centre and the interviews. As a result, they will fail.

The way to pass the selection process is to embark on a comprehensive period of intense preparation. I would urge you to use an action plan during your preparation. This will allow you to focus your mind on exactly what you need to do in order to pass. For example, if it has been many years since you last attended an interview, then you will probably have a lot of work to do in this area. If it has been many years since you last sat a test, then you may have to work very hard in order to pass the psychometric tests that form part of the assessment centre. The point I am making here, is that it is within your power to improve on your weak areas. If you use an action plan then you are far more likely to achieve your goals.

I use action plans in just about every element of my work. Action plans work simply because they focus your mind on what needs to be done. Once you have created your action plan, stick it in a prominent position such as your fridge door. This will act as a reminder of the work that you need to do in order to prepare properly for selection. Your action plan might look something like this:

My weekly action plan for preparing for Train Driver selection

Monday	Tuesday	Wednesday	Thursday	Friday
Research into the TOC I am applying for. Includes reading recruitment literature and visiting websites.	60 minute Interview preparation including preparing my responses to questions.	Obtain application form and read recruitment literature and application form guidance notes.	Research into the TOC I am applying for. Includes reading recruitment literature and visiting websites.	60 minute mock interview with a friend or relative.
60 minutes preparation for the Train Driver Tests.	30 minute Dots Concentration Test preparation.	45 minute Awareness and Recognition test preparation.	60 minutes preparation for the Train Driver Tests.	30 minute Dots Concentration Test.
20 minute jog or brisk walk.	30 minutes gym work.	20 minutes reading about the role of a Train Driver.	20 minute jog or brisk walk.	30 minutes gym work.

Note: Saturday and Sunday, rest days.

The above sample action plan is just a simple example of what you may wish to include. Your action plan will very much depend on your strengths and weaknesses. After reading this guide, decide which areas you need to work on and then add them to your action plan. Areas that you may wish to include in your action plan could be:

- Researching the role of a Train Driver;

- Researching the training that you will undergo as a Trainee Train Driver;

- Researching the Train Operating Company that you are applying for;

- Dedicating time to completing the application form and reading the guidance notes;

- Carrying out practice tests that are similar to the ones required at the assessment centre;

- Light fitness work in order to keep up your concentration levels;

- Interview preparation including carrying out a mock interview.

You will note that throughout the duration of this guide I make continued reference to 'preparation'. During my career I have been successful at over 95% of interviews and assessments that I've attended. The reason for this is simply because I always embark on a period of focused preparation, and I always aim to improve on my weak areas. Follow this simple process and you too can enjoy the same levels of success that I have enjoyed.

Finally, it is very important that you believe in your own abilities. It does not matter if you have no qualifications. It does not matter if you have no knowledge yet of the role of a Train Driver. What does matter is self belief, self discipline and a genuine desire to improve and become successful.

Enjoy reading the guide and then set out on a period of intense preparation!

Best wishes,

Richard McMunn

THE ROLE OF A TRAIN DRIVER

WHAT DOES A TRAIN DRIVER DO?

Before we start to take a look at the selection process for becoming a Trainee Train Driver, and more importantly how you can pass it, it is important that you learn about the role. As you can imagine many applicants get carried away with their childhood dreams and ambitions of becoming a Train Driver, without first establishing whether or not this is the job for them.

Train Drivers mainly operate either diesel or electric trains. There are strict rules on safety for both passenger and freight trains, which drivers must follow. In past years there have been a number of unfortunate rail incidents and Train Operating Companies want to employ people who take safety seriously. The word 'safety' should be at the forefront of your mind whilst you are going through selection. It's the exact reason why the TOC will assess your ability to concentrate for long periods of time during the assessment day. Because if you can't, then you will be putting yourself, your train and your passengers at risk from harm. During your initial training course you will receive much training and guidance on the important topic of safety. The role of a Train Driver does involve sitting down in the cab for prolonged periods of time. Before you apply you need to be certain that you are capable of maintaining high levels of concentration for lengthy periods.

Train Drivers also need to be familiar with their route, as during extreme weather conditions it can be difficult to negotiate the track. Therefore, it is important that you are capable of learning the route and being able to adapt quickly to a changing environment. During the selection process you will be assessed on your ability to learn and recall information, by way of assessing the key skills and abilities necessary for a job position as a Train Driver.

Many modern day trains will require you to make important

announcements to passengers as and when necessary and work can involve the opening and closing of the doors by a remote control system. Customer service is high on the agenda of the Train Operating Companies and therefore you should be able to provide examples of where you have already achieved this during a previous work related role. Being a Train Driver is not just about driving trains. It is about providing a high quality service to your customers, keeping them safe, and also keeping them updated on relevant travel/route information as and when the need arises.

As a Train Driver, you will need to have a detailed knowledge of the type of train you operate, the route that you are required to take, health and safety regulations, signalling systems and emergency procedures in the event of an incident.

TRAINEE TRAIN DRIVER TRAINING (Sample only)

Once successful, you will embark on a comprehensive training programme lasting approximately 12 months and your salary progression will be linked to this training. It is a demanding course, since each stage requires you to pass the relevant assessments before you can continue to the next level.

Stage 1 training

You'll spend two months learning about the 'cab environment' as well as the rules and regulations governing the safe operation of trains.

Stage 2 training

Now you're ready to start traction training, which will cover the appropriate electric multiple units depending on where you are based. This part of your training will last for approximately four months.

Stage 3 training

The next stage is the completion of 'route learning', which takes between two and three months. Once completed, you will be a fully qualified Train Driver!

WORKING HOURS AND CONDITIONS

Train Drivers normally work up to 37 hours a week. This will normally involve shift work and includes early starts, weekends, evenings and nights. Some long distance train drivers will stay overnight at the end of a run. Obviously there are restrictions on the number of hours you can work, and because of the high level of concentration required to drive a train, you will need to be at your best at all times.

As a Train Driver you will be supplied with a uniform by the Train Operating Company. On most journeys you will work alone unless you are being assessed by an inspector. The only other time that you will ride with another person is when a Trainee Train Driver is accompanying you. The Train Driver's working hours are spent sitting in the driver's cab. They cannot move around because they must concentrate on the signals and controlling the train. This includes monitoring the speed of the train and also keeping the 'dead man's handle' pressed down (this is a piece of safety equipment and if the driver lets it go, the train will stop automatically). Therefore, you should expect to work for long periods of time on your own, usually sitting down.

HISTORY OF TRAIN DRIVING

Despite decades of technological advances and the inexorable rise of the motor car, being a Train Driver is still the most popular dream job for many young boys. However, few youngsters actually get to pursue their childhood dreams when they get older, with most taking jobs in the administrative and support sectors.

Many people are now choosing Train Driving as a second career. As a result of this, a number of Train Operating Companies are seeing a big increase in the number of people applying to become a Train Driver later on in life. Another reason for this change in shift is due to the salary that a qualified Train Driver can earn. Depending on the company, Train Drivers can earn up to £37,000 per year which is a fantastic salary for most people! However, the money hasn't always been so good.

Train Driving stems back many, many years and the type and level of training provided used to be virtually non-existent. The long journey to becoming a Train Driver would start for most boys around the age of fifteen. They would work as engine cleaners and general assistants, often working for long hours until eventually, some years later, qualifying as Train Drivers. In the early years, there was no formal training and the only way to learn the trade of a Train Driver was to attend classes that were being run by experienced Train Drivers. These classes were known as Mutual Improvement Classes or MICs and they were attended by cleaners who were looking to progress up through the rosters to eventually become a qualified Train Driver. After a long period of time, cleaners would achieve the status of Fireman. On many occasions, Firemen had to work at depots that were many miles away from their home, but their will and desire to become a qualified Train Driver drove them on. Eventually, and after approximately eight years of hard study, they would be passed as a Train Driver. Unfortunately, this meant that they then had to start at the bottom of the ladder again and eventually work their way up to an Express Train Driver, often late on into their careers.

The situation remained the same until the 1960s, when Firemen were renamed as Secondmen or Drivers' Assistants. Eventually formal training was introduced and this was called MP12. The training ran for about six months in total and once a trainee had completed the course, he or she would then become a Relief Driver.

It wasn't until a Relief Driver reached the age of 21 that he or she would become a Train Driver. They would initially be posted out to unpopular depots that were quite a distance from home. They would then have to apply and be placed on a waiting list in order to move back closer to their home town.

HOW AND WHERE TO APPLY

Most TOCs do not advertise their vacancies publicly or in the local press/media. The reason for this is that they know they will get many applicants for their vacancies just by word of mouth or through waiting lists. Some TOCs provide an e-mail notification service that will inform you of any vacancies that become available. However, the best way to look for vacancies is to keep an eye on the TOC website that you are interested in joining. Simply go to one of the more popular search engines and type in the name of the Train operating Company that you want to work for. Then, simply keep a weekly check on the site for Trainee Train Driver vacancies. Within the 'Useful Contacts' section of this guide I have provided you with the contact details for each of the Train Operating Companies. You will also find a number of useful links to other websites that provide recruitment information.

In a nutshell, TOCs advertise vacancies as and when they become available. In terms of applying for vacancies many TOCs now allow you to apply on-line. However, if the option exists for you to send an application form directly in to them, then this should be your preferred route as it gives you the option of attaching a covering letter. See the Application Form section for more information.

BE THE FIRST TO BE NOTIFIED OF TRAIN DRIVER
JOBS!

At How2Become.com we do offer a Train Driver
job alerts service where you can sign up to be the first
to be notified about new Trainee Train Driver jobs.
Go to www.How2Become.com now and get yourself
on the list to be notified!

Due to the very nature that TOCs rarely advertise positions, I am aware that many people decide to join the TOC in an alternative role first. This could be as a Conductor, Guard, or even in the role of Customer Service staff. Whilst these roles may not be what you ideally want, they can give you a 'foot in the door' of the TOC. The benefits of joining a TOC in an alternative role first are plenty:

- If you are a good employee then the TOC are more likely to employ you as a Trainee Train Driver, given your previous good record;

- You will get to learn a lot about the TOC, how it operates, and you will also get the opportunity to build up some good contacts within the company;

- You will get the opportunity to speak to qualified Train Drivers. This will allow you to learn more about the role before you apply.

During the next section of the guide I will provide you with ten invaluable tips that will help you to pass the selection process.

TOP 10 TIPS AND ADVICE

TIP 1 – BE PATIENT AND PERSISTENT AND ALWAYS LOOK FOR WAYS TO IMPROVE ON YOUR WEAK AREAS

Many popular jobs, including the job of a Train Driver, are difficult to obtain. There are a number of reasons for this but the main one is the competition. There are approximately 400 applicants for every Trainee Train Driving position, so you need to stand out from the crowd in order to progress through each stage. Ask yourself how serious you are about becoming a Train Driver? Are you prepared to keep trying if you fail at one of the stages? Are you prepared to work as a Conductor first in order to get a foot on the ladder if you fail the selection process? Are you prepared to work hard to improve yourself and take the time required to prepare for the selection process?

The main thing to remember is that you can achieve your goal of becoming a Train Driver, providing you are prepared to work hard and improve. There are many ways of improving your chances of success and this guide will help you to understand them. I would strongly advise that you utilise an action plan to help you to concentrate on your weak areas. Whilst I have already covered this during the preface, it is worth mentioning it again. Within your action plan try to include lots of focus on your weak areas. Let's say for example you are not confident at your ability to show high levels of Awareness and Recognition (AART), a test that focuses on awareness and concentration. Within your action plan, you will need to write down exactly what you plan to do in order to improve your weak areas. This may include:

- Carrying out 20 minutes worth of timed tests five nights per week;

- Employing the services of a personal tutor to help you improve your concentration levels.

The key to improvement is to carry out what I call 'deliberate and repetitive' practice. This effectively means finding out exactly what you are weak at, and then carrying out lots of targeted practice until you can do it with your eyes closed! Not literally, but hopefully you get my point.

TIP 2 – LEARN ABOUT THE TRAIN OPERATING COMPANY YOU ARE APPLYING FOR

Put yourself in the shoes of the recruitment officer of the Train Operating Company you are applying to join. What type of person do you want to employ as a Trainee Train Driver? Someone who is serious about joining your TOC? Or someone who is just applying to any TOC with the sole purpose of securing a position as a Train Driver? The answer should be obvious. Therefore you need to learn about the TOC you are applying to join in order to increase your chances of success.

During my research into this guide I spoke to a number of recruitment officers from different Train Operating Companies and they all said the same thing – "You can spot a serial applicant a mile off!" A serial applicant is someone who applies for every Trainee Train Driver job regardless of the TOC. Whilst there is nothing wrong with this, there is a danger of that particular applicant not spending sufficient time on his or her application form and subsequently submitting a weak application. You may also find that a TOC will only allow you to apply for them if you reside within their area. Check with the TOC first to ensure you meet their minimum eligibility requirements.

When you complete the Application Form you will have the opportunity to make reference to the reasons why you want to become a Train Driver with that particular company. Throughout this guide I will show you how to increase your chances of success and demonstrate that you are serious about joining that company.

TIP 3 – PRACTISE PLENTY OF PSYCHOMETRIC TEST QUESTIONS

You'd be surprised at how many candidates attend the Assessment Centre having carried out little or no preparation whatsoever. Don't be one of those people who are ill-prepared.

The selection process involves a number of different tests, and we have provided practice tests that will help to improve the key skills and abilities, including:

- A Vigilance test;

- Checking and Assessing Information tests;

- A Concentration test;

- Awareness and Recognition tests;

- Co-ordination tests;

- Fast Reaction tests.

The majority of people who fail the assessment do so during the assessments of concentration. Concentration tests require high levels of concentration on repetitive and monotonous tasks, for prolonged periods of time. Although the test itself appears to be relatively simple, it is in fact difficult to master. Train Operating Companies need to be 100% sure that you can concentrate for long periods of time whilst carrying out repetitive tasks.

In addition to practising the sample tests that I have provided within this guide, I strongly encourage you to obtain additional testing books and resources that are available to you through www.How2Become.com.

Just by practising different psychometric tests your brain will begin to work quicker and more effectively and it is important that you take the time to do this. Make sure you allocate plenty of time in your action plan to each of the testing areas indicated above.

Please note:
The tests in this guide are not to be used as an indication of the actual tests. Instead, they should only be used as a guideline and preparation tool to practice the key skills and qualities required to become a Train Driver.

TIP 4 – CONSIDER ATTENDING AN ASSESSMENT DAY BEFORE YOU APPLY TO BECOME A TRAINEE TRAIN DRIVER

Part of the selection process for becoming a Trainee Train Driver involves attending an assessment centre. The assessment centre will include tests that assess similar skills to the tests that we have provided you with in this guide. You will be asked to sit a structured interview. The assessment centre is usually carried out by a qualified external organisation such as DB Schenker, although some TOCs do carry out their own 'in house' testing.

There is the option to attend an external based assessment day before you apply to become a Trainee Train Driver. Whilst this is not essential, it does have a number of key benefits:

- If you successfully pass the assessment day, you will have a certificate that is recognised by all TOCs;

- The validity of the certificate can last for up to 5 years, depending on the TOC you are applying for;

- If you already have the certificate when you apply to become a Trainee Train Driver, the TOC will not normally have to reassess you, saving them time and money.

The assessment day is suitable for candidates applying for the position of Trainee Train Driver with a Train or Freight Operating Company, fulfilling the requirements of the Railway Industry Standard 3751. Naturally, the assessment day comes at a cost and at the time of writing it is £130 with DB Schenker.

On passing the Train Drivers assessment you will receive your certificate. This will be valid for a period between 12 months and 5 years depending on the Train or Freight Operator you apply to join. The results are recognised by all Train Operating Companies. To find out more about the assessment day please visit the following website:

www.rail.dbschenker.co.uk

TIP 5 – SEND YOUR CV TO THE TRAIN OPERATING COMPANY THAT YOU WANT TO APPLY FOR

Trainee Train Driving vacancies are not a common sight. The reason for this is that most TOCs do not publicly advertise the vacancies. With this in mind I recommend that you send your CV to all the TOCs that you are interested in working for, along with a covering letter. Within your covering letter ask them to place you on their mailing list and request that they inform you of any future vacancies with their company.

Within this guide I have provided you with some essential tips and advice on how to create an effective CV and a sample covering letter. Take your time when constructing your covering letter and CV and take note of Tip 2! Your covering letter should include a brief amount of information about the company you wish to join and why you want to join them. Don't make the mistake of just sending out a 'blanket copy' letter and CV to all TOCs.

TIP 6 – UNDERSTAND THE IMPORTANCE OF BEING 'CUSTOMER FOCUSED'

All Train Operating Companies are competing to provide a high level of customer service. In any business it is the quality and reliability of service that will bring the customer back time and time again. Try to think of a shop or company that you have had a good experience with. What was so special about them that would make you go back to them? Most probably the level of customer service is the reason why.

When you apply to become a Trainee Train Driver you need to demonstrate a good 'customer-focused' attitude. You can demonstrate this both on your Application Form and during the Interview. TOCs want the people who work for them to provide the highest level of customer service possible so that the customer comes back to them time and time again. An unhappy customer is likely to tell ten other people about their bad experience, so the knock on effect can be huge. Think about the importance of being customer-focused and make it a prominent part of your application. If you have any previous experience of working in a customer-focused role then I advise that you include this on your application form.

TIP 7 – DEMONSTRATE AN AWARENESS OF SAFETY

Being safe and having the ability to follow safety rules and regulations is crucial to the role of the Train Driver. Safety must always come first! As a fully qualified Train Driver you have a lot of responsibility in relation to safety rules and regulations. If you cannot follow these rules and regulations, then you will not make a competent Train Driver.

Throughout the selection process you will be assessed against this important area and it is essential that you have it at the forefront of your mind at all times. When completing the Application Form you may be asked to provide an example of where you have been responsible for safety, or where safety has played an important part in a group or individual task. Think about your responses carefully and include answers that demonstrate your knowledge and understanding of the importance of safety. If you have any experience of carrying out a safety conscious role during current or previous employment then I would advise that you include this in your Application Form responses.

To learn more about safety in the workplace take a look at the Government's own Health and Safety website. This can be found at the following web address:

www.hse.gov.uk

TIP 8 – BE SMART AND PRESENTABLE

It goes without saying, but you'd be amazed at how many people turn up to the Assessment Centre, and even the Interview, in jeans and a t-shirt! Unsurprisingly, this does not go down well with the recruitment staff.

You are applying for a highly sought-after post with a Train Operating Company that most probably sets high standards. Therefore, it is important that you set yourself high standards right from the offset. Whenever you come into contact with the TOC or recruitment staff, you should always ensure that you are wearing formal clothing in the form of shirt and tie for men or smart dress for women. Make sure you are clean shaven and that your shoes are clean and tidy. You only get one chance to make a first impression so make sure it's a positive one.

Another piece of important advice is not to be late for any of your appointments. It is far better to arrive 15 minutes early than 1 minute late. There is a big emphasis by TOCs on ensuring their trains run on time. Therefore, they want to employ Train Drivers who are capable of getting to work on time.

TIP 9 – CONSIDER ANOTHER ROLE FIRST

Because the competition for becoming a Train Driver is so fierce, many people join a Train Operating Company in another role or position first in an attempt to get their foot in the door.

If there are no Trainee Train Driver vacancies with the Train Operating Company that you want to join then you may wish to consider becoming a Conductor or another similar role first, before eventually moving up to a Train Driver. This avenue certainly has its advantages. The main advantage is that the company will know how you work and what your attitude is like. Think about it carefully, if you were a Recruitment Officer for a Train Operating Company, who would you give the job to? A Conductor who had been working for you for a year and who was reliable and had a great track record, or someone whom you knew very little about? The answer is obvious; you would be more likely to give the job to the person you knew was reliable. Whilst this route is not essential it is certainly worth considering, especially if you have a long wait before the TOC are recruiting Trainee Train Drivers.

TIP 10 – TOC FIRST, TRAIN DRIVER SECOND

As you can imagine, many applicants are not too bothered which Train Operating Company they work for, so long as they get to become a Train Driver! In my view this is a mistake and it will not assist them in their application. Throughout the application process you should demonstrate that you have a desire to join that particular TOC, as opposed to being hell bent on simply becoming a Train Driver. The way to achieve this is to learn just as much about the TOC you are applying for as you do about the role of a Train Driver. During the interview there is a strong possibility that you will be asked questions similar to the following:

- Why do you want to work for our company?

- What research have you carried out into our company?

- What significant events have happened in this TOC over the past 12 months?

- What has attracted you to our company?

During your preparation for becoming a Trainee Train Driver make sure you carry out plenty of research into the TOC.

TRAINEE
TRAIN DRIVER
SELECTION
PROCESS

Every TOC is different, and will use a different selection process. However, there will be undoubtedly be similarities between the selection process for each TOC. An example of a Trainee Train Driver selection process is as follows:

The application form

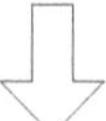

Psychometric testing/assessment centre

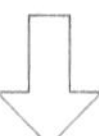

The structured interview + the managers interview

(Please note: sometimes the interviews can be conducted on the same day as the tests/assessments)

The Medical

The entire process can take many months and you will need a level of patience and determination if you are to succeed. Each of the above areas, with the exception of the medical, is covered in more detail during later chapters of this guide. Please remember to check with the Train Operating Company that you are applying to join that the above information is correct, as the selection process may vary.

THE APPLICATION FORM

Within this guide I have dedicated an entire section to the completion of the application form. The application form stage of the selection process is one of the hardest stages to get through. The reason for this is that there are approximately 400 applicants for every job. Therefore, it is essential that you take the time to complete the form correctly.

All Train Operating Company application forms vary. However, the general layout will take the following format:

- Information about the job you are applying for;

- Your contact information;

- About you;

- Reasons for your application;

- Your education, training and qualifications;

- Your employment details;

- Working time regulations;

- Your pastimes and other supporting information;

- Your health;

- Criminal record;

- Technical and professional questionnaire.

Whilst many of the sections on the application form are relatively straight forward to complete, it is still important that you take the time to complete each one of them accurately, ensuring your work is grammatically correct. Application forms either come in paper format or they are sometimes available to download and complete online through the TOCs website.

Before you complete your form make sure you visit the application form section of this guide for some useful tips and advice. The biggest mistake that most applicants make is the lack of attention to detail when completing the form. Many people apply for a number of vacancies all at once and tend to 'cut and paste' their responses into every form. This is not the ideal approach to take and one that I would recommend avoiding. The TOC want to know that you are interested in joining their company and they do not want to employ people who are merely set on becoming a Train Driver regardless of the TOC.

PSYCHOMETRIC TESTING

Psychometric tests form an integral part of the selection process and are normally split over four different areas. Within this guide, we have done our utmost to ensure you practice tests that assess the same skills and attributes required for the real tests that you will need to sit in your assessment.

The majority of TOCs will employ a specialist independent company to conduct the assessments on their behalf. There are currently only a few Assessment Centres that have the capability of carrying out such assessments and these are based at Doncaster, Crewe and London. Before you commence each of the tests you will have the opportunity to carry out a number of practice sample questions. Obviously the best way to prepare is to carry out lots of practice tests well in advance of your assessment day. You should start the practice questions at least two weeks before your assessment date so that your mind gets used to working under pressurised conditions.

Remember that the competition is fierce and the more you prepare the greater chance of success you will have.

You may decide to pay for an assessment yourself in order to gain a certificate that is recognised by all Train Operating Companies. The cost of this can vary but it is usually around £130 and, if your budget can stretch this far, I recommend you take the opportunity. Visit the following website for more details:

www.rail.dbschenker.co.uk

Whilst this book contains sample questions regarding the train driver selection process, you should note that these tests should only be used as a way of practising your training; they are not a real depiction of the actual tests. We have provided these questions in order to allow you to familiarise yourself with the key skills and attributes that TOCs are looking for.

HOW TO CREATE AN EFFECTIVE CV

As I discussed earlier in the guide it is advisable to send your CV and a covering letter to the TOCs you are interested in applying for. During this section of the guide I will provide you with a step by step guide on how to create a CV that is both effective and relevant to the TOC you want to join.

The word Curriculum Vitae translated means the 'course of life'. CV's are used to demonstrate to an employer that you have the potential, the skills, and the experience to carry out the role you are applying for. Your CV is a very important document and you should spend sufficient time designing it so that it matches the job that you are applying for as closely as possible.

What Makes an Effective CV?

In simple terms an effective CV is one that matches the specification and the requirements of the job you are applying for. Your CV should be used as a tool to assist you in your application for becoming a Trainee Train Driver and it should be centred on the following areas:

- Creating the right impression of yourself;

- Indicating that you possess the right qualities and attributes to perform the role of a Trainee Train Driver;

- Grabbing the Train Operating Company's attention;

- Being concise and clear.

The most effective CV's are the ones that make the assessor's job easy. They are simple to read, to the point, relevant and focus on the job/role that you are applying for. CV's should not be overly long unless an employer specifically asks for this. Effective CV writing is an acquired skill that can be obtained relatively quickly with a little bit of time, effort and focus.

Before you begin to start work on your CV it is a good idea to have a basic idea of how a job/person specification is constructed. A job description/person specification is basically a blueprint for the role you are applying for; it sets out what the employer expects from potential applicants. One of your main focus points during the construction of your CV will be to match the job/person specification of the Trainee Train Driver. Most job/person specifications will include the following areas:

EXPERIENCE REQUIRED: previous jobs, unpaid work experience, life experience, skills, knowledge and abilities: for example, languages, driving, knowledge of specialist fields, ability to use equipment, plus some indication of the level of competence required, and whether the person must have the skills or knowledge beforehand or can learn them on the job.

QUALIFICATIONS REQUIRED: exams, certificates, degrees, diplomas (some jobs require specific qualifications, but most do not and it can be fairer to ask for the skills or knowledge represented by the qualification rather than asking for the qualification itself).

PERSONAL ATTRIBUTES REQUIRED: such as ability to concentrate, a willingness to work on one's own and the ability to work within a team environment.

PERSONAL CIRCUMSTANCES: such as residing in a certain area and being able to work weekends or evenings.

Most job/person specifications will be based around a task analysis of the vacancy, so there should be nothing within the job description/person specification that is irrelevant or that does not concern the particular role you are applying for. Whatever requirements you are asked to meet, you should try hard to match them as closely as possible, providing evidence if possible of your previous experience.

What is the TOC looking for in your CV?

As previously stated you should ensure that you make the assessor's job as simple as possible. Try to put yourself in the shoes of the assessor. How would you want an applicants CV to look? You would want it to be relevant to role they are applying for and you would want it to be neat, concise and well organised. You need to spend some time thinking about the type of person the TOC are looking for and how you can match the specification that is relevant to the job you want. Most job specifications will list the essential/desirable requirements in terms of education, qualifications, training, experience, skills, personality and any other special requirements.

Let's take a look at some of the skills and qualifications required to become a Trainee Train Driver. Please note, the following details are not relevant to any particular TOC and you may find that requirements vary for each different company.

Qualifications required	About the job
Essential: • Good Level of Secondary Education. • Basic Math/ English. Desirable: • O level GCSE Standards Math/ English.	• To ensure our trains are driven in a safe, punctual and economical manner, in accordance with the Rules, Regulations, Appendices and Publications currently in force for this Train Operating Company. • Responsible for booking on & off in person at the train station. • Whilst on duty wear the required uniform or safety clothing and maintain such clothing to the required TOC standard. Whilst on duty conduct self to the standards of behaviour required by the business.

- Responsible before and during commencement of duties for reading of all notices, speed restrictions and alterations to diagrams affecting self and routes covered during the course of driving duties. Sign for and read all Weekly Special notices, Periodical Notices and amendments to Rules & Regulations.

- Take charge of trains at commencement of duty and following breaks in sufficient time to allow for punctual departure. Make every effort to maintain punctuality of services within the constraints of speed restrictions, signalling and weather conditions.

- Responsible for informing the Operations Delivery Centre / Rolling Stock Technician at the earliest opportunity, of any faults or failures to TOC stock that cannot be rectified following reference to faults & failures manuals.

- Maintain knowledge of Routes, Permanent Speeds, Signals, Sidings, Depots and Traction required by the business. Notify Driver Manager of any deficiencies in your knowledge of such items at the earliest opportunity.

You will see from the above details that some of the key elements of the role include an ability to read and follow rules and procedures, a level of self discipline, and knowledge of routes and regulations. Once you have the above information then you will be able to mould your CV around the key aspects of the job.

Before I provide you with a sample CV that is based on matching the above role, let's first of all take a look at some of the key elements of a CV.

The key elements of a CV

The following is a list of information I recommend you include within your CV. Try to put them in this order and remember to be brief and to the point. Make sure you include and highlight the positive aspects of your experience and achievements.

- Your personal details;

- Your profile;

- Your employment history;

- Your academic acievements;

- Your interests;

- Any other information;

- Your references.

Let's now take a look at each of the above sections and what you need to include.

YOUR PERSONAL DETAILS

When completing this section you should include the following details:

- Your full name

- Address

- Date of birth

- Nationality

- Contact telephone numbers including home and mobile

- E-mail address

YOUR PROFILE

To begin with try to write a brief but to the point statement about yourself making sure you include the keywords that best describe your character. Some effective words to use when describing yourself might include:

Ambitious, enthusiastic, safety conscious, customer focused, motivated, caring, trustworthy, meticulous, sense of humour, drive, character, determination, will to succeed, passionate, loyal, teamwork, hard working.

The above words are all powerful and positive aspects of an individual's character. Try to think of your own character and what positive words you can use that best describe you.

Within your profile description try to include a statement that is relative to you and that will make the TOC think you are the right person for the job, such as:

"I am an extremely conscientious and hard working person who

has a great deal of experience in Health and Safety. I have very good organisational and motivational skills and I am always striving to improve myself. I believe that I would embrace the challenges that this new role has to offer and I am able to learn large amounts of job relevant information and procedures."

YOUR EMPLOYMENT HISTORY

When completing this section try to ensure that it is completed in reverse chronological order. Provide the reader with dates, locations and employers, and remember to include your job title. Give a brief description of your main achievements and try to include words of a positive nature, such as:

Achieved, developed, progressed, managed, created, succeeded, devised, drove, expanded, directed.

It is also a good idea to quantify your main achievements, such as:

"During my time with this employer I was responsible for carrying out difficult tasks whilst under pressure."

YOUR ACADEMIC ACHIEVEMENTS

When completing this section, include the dates, names and locations of the schools, colleges or universities that you attended in chronological order.

You should also include your qualifications and any other relevant achievements such as health and safety qualifications or first aid qualifications. Anything that is relevant to the role you're applying for would be an advantage.

YOUR INTERESTS

Within this section try to include interests that match the requirements of the job and ones that also portray you in a positive manner. Maybe you have worked within the voluntary sector or have even carried out some charity work in the past? If so try to include these in your CV as they show you have a caring and concerning nature. In relation to the role of a Trainee Train Driver, the following activities and past times are recommended:

- Playing a team sport or activity. This demonstrates you have the ability to work with others as part of a team.

- Recent study activities such as learning a new qualification. This demonstrates that you are able to learn and retain a large amount of job specific information.

- Playing a musical instrument. This demonstrates you have the ability to learn something new, and that you have the patience and determination to succeed.

ANY OTHER INFORMATION

Within this section of your CV you can include any other information that is relevant to your skills or experiences that you may feel are of benefit. Examples of these could be certificates of achievement from school or work.

REFERENCES

Within this section try to include your current or previous employer, providing you know that they are going to write positive things about you. Be careful who you choose as a reference and make sure you seek their permission first prior to putting down their name and contact details. It may also be a good idea to ask them if you can have a copy of what they have written about you for reference later.

Sample CV

The following sample CV has been designed to give you an idea of how an effective CV might look. It has been created with the position of Trainee Train Driver in mind. All of the information provided is fictitious.

Curriculum Vitae of
Richard McMunn

Address: 75, Any Street, Anytown, Anyshire. ANY 123
Date of birth: 01/01/1970
Nationality: British
Telephone contact: 01227 XXXXX / Mobile 07890 XXX XXX
E-Mail contact: richardmcmunn@anyemailaddress.co.uk

Personal Profile of Richard McMunn

I am an extremely fit and active person who has a great passion for the Rail industry and I have a track record of high achievement. I have very good organisational and motivational skills and I am always striving to improve myself. I believe that I would embrace the challenges that this new role has to offer. I am a motivated, dedicated, loyal and ambitious person who has the ability to work both within a team and also unsupervised.

I already have a large amount of experience in the working environment and take on a large number of responsibilities both at work, around the home and in my leisure time activities. I currently hold a Health and Safety qualification and I am fully aware of the importance of safety in the role that I am applying for. I have experience in a customer service environment and have the ability to act as a role model for the Train Operating Company. I understand that the role of a Train Driver does not just involve driving trains but it also means acting as a point of contact for the valued customers and providing them with an excellent service at all times.

To conclude, I am a fit, motivated, active, organised and professional individual who has a lot of skills and experiences to offer this Train Operating Company.

Employment history of Richard McMunn (in chronological order)

Job position/title/company #1 goes here
Date of employment goes here
During my time with this employer I was responsible for motivating my team and organising different activities.

Job position/title/company #2 goes here
Date of employment goes here
During my time with this employer, I was responsible for stock taking and dealing with customer's queries and complaints. I also took on the responsibility of arranging the company's annual staff leisure activity event which often included some form of motivational talk.

Job position/title/company #3 goes here
Date of employment goes here
During my time with this employer I undertook a training course in health and safety and first aid. Part of my role included managing resources and carrying out risk assessments as and when required.

Academic achievements of Richard McMunn

Health and Safety qualification	Date of achievement goes here
First Aid qualification	Date of achievement goes here
GSCE Maths Grade C	Date of achievement goes here
GCSE English Grade C	Date of achievement goes here
GCSE Physical Education Grade B	Date of achievement goes here

Interests and Hobbies of Richard McMunn

I am an extremely fit and active person who carries out a structured training programme at my local gym five times a week. During my training sessions I will carry out a variety of different exercises such as indoor rowing, cycling, treadmill work and light weights. I measure my fitness levels by performing the multi-stage fitness test once a week and I can currently achieve level 14.5. In addition to my gym work I am a keen swimmer and break up my gym sessions with long swim sessions twice a week. I can swim 60 lengths of my local swimming pool in time of 35 minutes.

I am also the Captain of my local football team and play in the position of midfield. I am also responsible for organising and arranging the weekly training sessions.

In addition to my sporting activities I like to relax with a weekly Yoga group at my local community centre. I also have a keen interest in art and attend evening classes during the months October through to December.

Further Information

Six months ago I decided to carry out a sponsored fitness event in order to raise money for a local charity. I swam 60 lengths of my local swimming pool, and then ran 26 miles before cycling 110 miles all in one day. In total I managed to raise over £10,000 for charity.

References

Name, address and contact details of reference #1
Name, address and contact details of reference #2

TOP TIPS FOR CREATING AN EFFECTIVE CV

New application = new CV

It is important that every time you apply for a job you re-evaluate the content of your CV so that you can match the skills and qualifications required. As a rule you should complete a new CV for every job application unless your applications are close together and the job/person specification is relatively the same. Don't become complacent or allow your CV to get out of date.

Don't pad out your CV

There is a common misconception amongst many job applicants that you need to make your CV scores of pages long for it to get recognised. This simply isn't true. When creating your CV aim for quality rather than quantity. If I was looking through an applicants CV then I would much prefer to see three to five pages of high quality focused information rather than thirty pages padded out with irrelevance.

Create a positive image

Writing an effective CV involves a number of important aspects. One of those is the manner in which you present your CV. When developing your CV ask yourself the following questions:

- Is the spelling, grammar and punctuation correct?

- Is it legible and easy to read?

- Is the style in which you are writing your CV standardised?

- Is it neat?

- Is it constructed in a logical manner?

By following the above tips in respect of your CV, you will be on the right track, excuse the pun, to improving your chances of getting a job as a Trainee Train Driver. You should spend just as much time on the **presentation** of your CV as you do on the content.

Do you have the right qualities and attributes for the job you are applying for?

When you are developing your CV have a look at the required personal qualities that are listed within the job/person spec. Try to match these as closely as possible but again, ensure that you provide examples where appropriate. For example, in the sample job description for a Trainee Train Driver one of the required personal qualities was:

'To ensure our trains are driven in a safe, punctual and economical manner, in accordance with the Rules, Regulations, Appendices and Publications currently in force for this Train Operating Company.'

Try and provide an example of where you have achieved this in any previous roles. The following is a fictitious example of how this might be achieved:

"In my current role as a delivery driver I am responsible for ensuring that all deliveries arrive on time and in a punctual manner. It is also my responsibility to check the safety of the vehicle every morning before I start work and to stick to the speed limits and rules of the Highway Code. I always ensure that I follow company rules and procedures in relation to working practices."

Matching your qualities and attributes to the role you are applying for is very important.

Be honest when creating your CV

If you lie on your CV, especially when it comes to academic qualifications or experience, you will almost certainly get caught out at some point in the future. Maybe not straight away but even a few months or years down the line an employer can still dismiss you for incorrect information that you provide during the selection process. It simply isn't worth it. Be honest when creating your CV and if you don't have the right skills for the job you are applying for, then go out there and get them!

Now that I've shown you how to create an effective CV, schedule into your action plan a date and time when you intend to create your own. Now let's move on to how to complete the application form correctly.

HOW TO COMPLETE THE APPLICATION FORM

THE APPLICATION FORM

Train Operating Company application forms vary. However, the types of questions that you will be asked to respond to are relatively similar in nature. In this section of the guide I have provided you with hints, tips and advice on how to increase your chances of progressing through this important stage. Yes it is true that only a small percentage of people make it through this initial stage of the selection process, but this doesn't mean you have little or no chance of succeeding. Follow the guidance that I have provided you within this section and your chances will greatly increase.

THE COVERING LETTER

Whether you are applying by CV or Application Form you should also enclose a covering letter. However, do bear in mind that a lot of train operating companies accept online applications only. Sending a covering letter shows the TOC that you are serious about joining their particular organisation. The manner in which you formulate your covering letter is very important. If you simply create a generic covering letter and then send it to lots of different companies, the TOC will see right through it. When constructing your letter (see example provided) make it presentable, logical and relevant to the TOC you are applying to join. I also advise that you personally write the letter as this will demonstrate to the TOC that you have taken more time and effort in your application. Remember to state what job you are applying for and remember that letters that are seeking 'any job' are more likely to be rejected.

Within your letter state the reasons why you want to join this particular TOC and demonstrate that you know a little bit about their company. Once again this will add some weight to your application. Draw attention to one or two points expanded on in your CV or application form to explain why you think you are well suited to the job with this company. Before sending off your application, try to find out the name of the person who will be on the receiving end of the letter. This can be achieved by contacting the TOC Recruitment Office either by e-mail or telephone (see Useful Contacts section). If the letter is addressed to them personally, then again, this will show how serious you are.

Finally, as with all letters, you need to check for correct grammar, spelling and punctuation. You may wish to get a friend to help you read through the letter and check for any errors. Take a look at the following sample covering letter. This will give you some ideas on how to construct your own letter based on your own individual circumstances.

SAMPLE COVERING LETTER

Mr Richard McMunn,
31, Fictitious Street,
Fictown,
Ficton FCT 1AW

Dear Sir/Madam,

RE: Trainee Train Driver position reference 126/TTD

Please find enclosed my completed application form and CV pertaining to the above position. I am applying for this post because I see Ficshire Rail as an exciting and forward-thinking company and I would very much like to work for you. In particular, the position of Trainee Train Driver is one that I have been working towards for many months now. In my quest to become a Train Driver I have researched your company thoroughly and have been impressed with the 'Meet The Managers' section on your website. This shows that the company is serious about its customer relations and I believe that I have the professionalism and high standards that are required to work with your company and can also meet the expectations you require of all employees.

I am a highly ambitious, safety conscious and reliable team player who can be trusted to carry out pressurised tasks. I would like to take this opportunity to thank you for reading my application and CV and if you need any more information from me then please let me know so that I can forward it on to you as soon as possible.

Yours faithfully,

Richard McMunn

PREPARING TO COMPLETE THE APPLICATION FORM

Most of the sections on the application form are relatively straightforward to complete. However, there are a number of very important sections that will need your utmost attention if you are to succeed. First of all, read the following tips and advice that relate to the completion of your application form:

Read Everything First

This applies to both the application form, the accompanying literature (if any) and the TOC website. You will need to understand a little bit about the company first before you can successfully complete the application form. You should also study the job description, person specification and the accompanying recruitment guidance notes.

Correct Ink Colour

Unless you are submitting an on-line application make sure you read any requirements that relate to ink colour or capital letters etc. The TOC Recruitment Office will receive hundreds of applications for every job advertised and the initial sift will look at minor errors such as these. If you cannot follow simple instructions such as the correct ink colour then there is not much chance you'll be able to operate a train safely. Read everything carefully and follow all of the instructions.

Complete a Rough Draft First

The first time around you are more than likely to make some mistakes. I advise that you photocopy the application form first (unless you are completing an on-line version) and complete a rough draft first. This will give you the opportunity to practice. Then, once you have finished your application, take a copy of it so that you can refer to it before you attend the interview. The interview panel will most certainly refer to your application form during the Manager interview.

On the following pages I have provided you with a number of sample responses to some of the more common types of application form question. The 'question and answer' sections on the application form are very important and represent an opportunity for you to show the recruitment staff at the TOC how good you are. Before each question I have explained what the question means and how best to construct your response to it. Then, I have provided you with a sample response to each question. Please note that these are to be used as a guide only. It is important that you answer the questions on your application form based on your own experiences and knowledge.

SAMPLE QUESTION NUMBER 1

Now that you've read more about the job, please tell us why you're applying for it and what knowledge, experience or skills you have that might be relevant.

The clue in this type of question is to READ about the job you are applying for. The question is asking you to match your knowledge, experience and skills with the job you have applied for. Therefore you need to read the job description before responding. Job descriptions or person specifications usually have both 'essential' and 'desirable' criteria included. Basically you must provide evidence of where you can meet the 'essential' criteria on your application form. Desirable criteria will gain you extra marks but they are not essential.

If the TOC have not sent you a copy of the job description then try to obtain a copy of it before completing the form. This will give you an insight into the role that you are applying for. Once you have read the information about the post you will then be able to construct a suitable answer. Try to include any knowledge, skills or experience you may have that relates to the job description. If you have experience or knowledge in health and safety, working in pressurised situations or working in a customer-based environment then you may wish to include these also.

The Trainee Train Driver position is one that requires your 100% attention at all times, and an ability to learn new skills quickly and accurately. Again, if you have experience of working in these conditions then say so. Now take a look at the following sample response before constructing your own, based on your own skills, knowledge and experience.

SAMPLE RESPONSE TO QUESTION NUMBER 1

I am applying for this post because I am looking for a new and challenging role. I enjoy working in a customer-focused environment and believe I would make an excellent Train Driver for your company. I am also prepared to relocate to live within 30 minutes of the depot. I understand that the company is changing and moving forward and I believe you would be an exciting company to work for. I also believe I can bring something to the team in terms of commitment, motivation and enthusiasm.

I have worked in a customer-based role for a number of years now and during this time I have developed skills that can be applied to the role of a Trainee Train Driver. As well as being a good communicator and possessing excellent practical skills I am also highly safety conscious and understand that this is a very important element of the role. In addition to my 12 years experience in a customer-focused role I worked for 4 years with the Royal Navy. I am therefore a highly disciplined person and a very good team player. I have educational qualifications in English Language, English Literature and Art and I am also coming to the end of studying for a Diploma in Management Studies. I also hold a Health and Safety qualification through IODA in Nottingham. I am a fit and active person who visits the gym/swimming pool three times a week and I also play football for a local Sunday team. I am a very good communicator and learn new skills quickly. I am used to working long and varied hours and I understand that the role requires a high level of flexibility, which I am prepared for. I enjoy working with, and meeting people from all walks of life and I truly value the benefits of a diverse workforce. To summarise, I am a highly professional, caring, trustworthy, friendly and motivated person and I believe I would make an excellent member of the Train Operating Company team.

SAMPLE QUESTION NUMBER 2

Please tell us about anything you get up to outside work that gives us a better idea of what you're like as a person and why you might be right for our company.

This type of question is designed to assess the type of person you are outside of work. This will give the company an idea of how you are likely to perform at work and will tell them if you are fit, healthy and active. When responding to this type of question, make sure you make reference to the job description. What type of duties will you be required to perform and can you match your external activities to them? Being fit and active is always a positive aspect that the recruitment staff will be looking for. If you are active outside of work, then you are also likely to be active at work and achieve your tasks to the required standard. If you have recently achieved any educational or academic qualifications outside of work then it would be a good idea to make reference to these too. Now take a look at the sample response before creating your own based around your own skills, knowledge and experience.

SAMPLE RESPONSE TO QUESTION NUMBER 2

KEEPING FIT – I attend the gym at least 3 times per week and carry out some light weight work. Whilst at the gym, I usually perform 20 minutes of rowing each time and cover a distance of 5,000 metres. I particularly enjoy swimming and swim 50 lengths, 3 times per week. When I get the opportunity, I like to go walking, too, in order to keep healthy. Staying fit and healthy means that I am able to maintain a high level of concentration at work and it also helps to keep my enthusiasm and motivation levels high. This shows that I am a dedicated and determined person who is always looking to improve himself.

MUSICIAN – I currently play the drums and piano. I have always enjoyed being creative and I play the drums in a function band that plays at wedding events and parties on some weekends. This shows that I have the dedication to learn new skills and I have the ability to concentrate on the task in hand when required. Learning new skills is essential to the role of a Trainee Train Driver and I believe that I have the ability to learn new skills quickly and adapt them to the work environment in a safe and effective manner.

SAMPLE QUESTION NUMBER 3

As the role you've applied for means that you'll be dealing with the safety of our customers and the delivery of our operation, we would like to hear examples of how you have used your initiative to solve a difficult problem.

Having the initiative to solve problems is integral to the Train Driver's role. On many occasions you will be on your own for many hours and you will have many responsibilities. Whilst you will have set procedures and guidelines to adhere to, you must still have the required initiative to solve difficult problems. Before responding to questions of this nature, make sure you READ the question first and try to understand what is required. Remember to write a response that identifies the use of your initiative to solve a difficult problem.

SAMPLE RESPONSE TO QUESTION NUMBER 3

During a recent staff meeting I was aware that there were a number of problems between some members of the team. The team wasn't working effectively so we all discussed ways in which we could improve. The actions of the team were starting to have an effect on the team's performance, so I decided to take the initiative to resolve the issue. I facilitated the meeting and asked everybody to share their views and opinions. I listened to each person individually and tried to encourage people to come up with solutions in order to improve the team's effectiveness. A positive point that came from our discussions was that people felt that we didn't hold enough meetings to talk about the problems we all face. It was agreed that with immediate effect we would hold weekly meetings to discuss issues, gather and share information, and look for ways that we could all support each other in our work. Since the meeting the team has moved forward and is now working far more effectively.

SAMPLE QUESTION NUMBER 4

As the role you've applied for means that you'll be dealing with the safety of our customers and the delivery of our operation, we would like to hear examples of how you have developed your abilities to improve yourself.

Having the ability to constantly review your own performance and take steps to improve is an important aspect of everyday life. This is particularly relevant in the workplace and, with a role that requires high safety standards and adherence, this is very much the case.

When responding to this type of question, try to think of an example or examples where you have improved yourself. This may be through a training course or educational qualification(s). You must ensure that you provide good examples to this type of question as the TOC recruitment staff want to see that you have the ability to pass the initial training course and absorb new skills relatively easily. Now take a look at the following sample response before creating your own.

SAMPLE RESPONSE TO QUESTION NUMBER 4

In order to carry out my duties in my current role effectively, I felt that I needed more management skills. I decided to pay for, and embark on, a Diploma Course in Management. I am coming to the end of the course and have found it a useful tool for improving my skills. I am always looking for new ways to improve my skills and knowledge so that I can perform better both in a professional and personal capacity. I also believe it is important to keep fully up-to-date and conversant with company health and safety policies and, every week, I read the company safety log to ensure I am aware of any changes or amendments to the policy. The safety of our customers and clients is paramount and I have developed many skills in this area that I believe would be an asset to the role of Trainee Train Driver with your company.

SAMPLE QUESTION NUMBER 5

As the role you've applied for means that you'll be dealing with the safety of our customers and the delivery of our operation, we would like to hear examples of how you have played a positive role as team member or leader.

Having the ability to work as an effective team member is important in any organisation and Train Operating Companies are no exception. The TOC will be made up of many different people, all of whom have an important role to perform. Therefore, it is essential that you have had some experience of working in a team environment, either as a team member or team leader. Try to think of an occasion when you have been part of a team or have even been the leader of a team. When responding to questions of this nature, think of a scenario where you worked as part of the team to achieve a task or solve a problem. Now take a look at the following sample response before using a blank sheet of paper to construct your own.

SAMPLE RESPONSE TO QUESTION NUMBER 5

In my current role, I am responsible for the safety of my team and for ensuring that any health and safety incidents are reported in line with company regulations. I am also involved in coaching and mentoring my team and providing them with feedback, often helping them to improve. I currently lead a team of 18 staff and I am required to ensure the team operates effectively in terms of management, health and safety, and training. Following any incident that relates to health and safety I always fully brief each member of the team to ensure that I have done everything in my power to prevent an incident occurring again.

SAMPLE QUESTION NUMBER 6

As the role you've applied for means that you'll be dealing with the safety of our customers and the delivery of our operation, we would like to hear examples of how you have had to work under pressure.

If you are successful in your pursuit of becoming a Train Driver, you will undoubtedly have to work under pressure on some occasions. Maybe you will experience technical difficulties whilst on the train or the air conditioning will fail. You will have to remain calm and focused and this question is designed to assess your ability to do just that. Try to think of a scenario where you have worked under pressure but still achieved the task or goal.

Take a look at the following sample response before using a blank sheet of paper to construct your own response based on your own experiences.

SAMPLE RESPONSE TO QUESTION NUMBER 6

In my current role as customer service manager I am required to work under pressure on a daily basis. Recently, I was presented with a situation where two members of staff had gone sick leaving me with only three other staff members to manage the shop during a busy Saturday.

During the morning we were due to take a stock delivery which meant that I had to perform many tasks without taking a break. During the day I dealt with two customer complaints, took delivery of the stock, served customers whilst others took their break and also dealt with a fire alarm actuation. I am often required to perform under pressure and thrive in such conditions. I always adapt well to situations like these and ensure that I still maintain a high level of professionalism at all times.

SAMPLE QUESTION NUMBER 7

As the role you've applied for means that you'll be dealing with the safety of our customers and the delivery of our operation, we would like to hear examples of how you have taken responsibility to communicate an important message.

As a Train Driver you will have to communicate important messages to the customers. The messages may relate to the delayed departure or arrival of the train and these must be communicated sensitively. Try to think of an occasion where you have had to communicate an important message where you were under pressure. Take a look at the following sample response which will help you to create your own. Once you have read the provided example, use a blank sheet of paper to construct your own response based on your own experiences.

SAMPLE RESPONSE TO QUESTION NUMBER 7

Whilst working in my current position as a sales person I was the duty manager for the day as my manager had gone sick. It was the week before Christmas and the shop was very busy. During the day the fire alarm went off and I started to ask everybody to evacuate the shop, which is our company policy. The alarm has gone off in the past but the normal manager usually lets people stay in the shop whilst he finds out if it's a false alarm. This was a difficult situation because the shop was very busy, nobody wanted to leave, and my shop assistants were disagreeing with me in my decision to evacuate the shop. Some of the customers were becoming irate as they were in the changing rooms at the time. Both the customers and my shop assistants were disagreeing with me. The customers were saying that it was appalling that they had to evacuate the shop and that they would complain to the head office about it. My sales staff were trying to persuade me to keep everybody inside the shop and that it was most probably a false alarm, like it usually is. I was determined to evacuate everybody from the shop for safety reasons and would not allow anybody to deter me from my aim. The safety of my staff and customers was at the forefront of my mind, even though it wasn't at theirs. I persisted with my actions and eventually got everybody to leave the shop. When the Fire Service arrived they informed me that there had been a small fire at the rear of the shop and that the actions I had taken were the right ones. Everybody was safe and nobody was hurt as a result of the incident.

FINAL TIPS FOR CREATING A SUCCESSFUL APPLICATION FORM

- Read the form carefully before starting to complete it. Also be sure to read all of the accompanying guidance notes, person specification and job description.

- Follow all instructions carefully. Your form can be rejected for failing to follow simple instructions.

- If you are completing a handwritten version of the form make sure your handwriting is neat, legible, concise and grammatically correct. You will lose marks for incorrect spelling!

- Before you submit the form get somebody to check it over for you.

- Once you have completed the form make sure you make a photocopy of it. You will be asked questions that relate to your responses during the Manager interview.

- Send the form recorded delivery. I have known of many application forms to go missing in the post.

THE TRAIN DRIVER TESTS

Before I provide details about the practice tests and how they test the same skills and qualities of your real test, I want to explain a little bit more about how you will be assessed during the selection process.

Basically, the assessors are looking for a number of key qualities and attributes to be demonstrated. The following pages will provide you with excellent 'insider' information on what they are looking for. All you have to do, is demonstrate the requirements of each assessable area in order to improve your chances of success.

The 'assessable qualities' are listed on the next few pages, followed by a description of each requirement and how you must perform.

You should be aware that the tests that are provided in this book are purely for you to practice the key skills and attributes required of any successful Train Driver. So, you should not use these tests as insight into the actual tests. Instead, you should use these practice questions to practice the key skills and qualities that are going to be assessed.

THE TRAIN DRIVER ASSESSABLE QUALITIES AND REQUIREMENTS

Attention

Selective attention (including perceptual differentiation)

Ability to differentiate between different sources of information and attend selectively to them, e.g. distinguishing and attending to alarms.

Attention switching

Ability to switch attention between sources of information, e.g. lineside information and in-cab information.

Vigilance

Sustained attention; Concentration

Ability to receive and manage selected stimuli under time pressure and maintain attention over an extended period of time.

Decision making

Multi-tasking

Ability to perform different tasks in parallel, eg making train announcements while on the move.

Prioritisation

Ability to prioritise information and tasks.

Planning and anticipation

Ability to interpret information and plan for future events, e.g. identifying a landmark cue before a station and start to decelerate.

Fault diagnosis

Ability to interpret information from a variety of sources, and use this to recognise, diagnose, and report on faults.

Instrument interpretation

Ability to understand and interpret information from instrumentation.

Trainability

Ability to retain, recall and apply job related information

Ability to learn new information in appropriate time limits, e.g. a new temporary speed restriction.

Ability to learn new information in appropriate time limits

Ability to retain, recall, and apply job related information, e.g. specific rules or procedures.

Openness to experience

Willingness to try new approaches, learn from mistakes, and change working practices as required.

Communication

Spoken communication

Ability to effectively communicate information verbally, including using verbal protocols to convey safety critical communication.

Written communication

Ability to communicate effectively in writing.

Reading

Ability to read and understand information, and respond appropriately, e.g. Weekly Operating Notices (WONs).

Listening

Ability to listen to, and understand information, and respond appropriately, e.g. instructions from a signaller.

Speed of reaction

Quick and adequate response to simple and complex visual and acoustic stimuli and the associated quality of performance.

Controlled motor coordination

Appropriate and controlled movements in response to decisions about complex stimuli.

Ability to anticipate elements in a traffic environment and make a correct decision about how to respond given the speed and distances involved.

Self-control

Calmness under pressure

In pressured situations, the extent to which the candidate maintains an appearance of being calm and in control.

Reactivity to stress

The extent to which the candidate maintains effective performance under pressure (in terms of quality and rational decisions/ actions).

Assertiveness

The extent to which the candidate is confident, effective and direct in communicating with others and maintains an independent position.

Rule compliance

Compliance with rules and procedures

Willingness to comply with formal and informal rules and procedures that self or others are expected to follow for safety or legal reasons.

Conscientiousness

Dependability

Extent to which the candidate can be relied upon to undertake tasks and fulfil responsibilities.

Attitude to work and people

Willingness to accommodate, cooperate, and help colleagues at work and to comply with own and other's work-based rules.

Commitment to work

Extent to which candidate does a job well and to best of ability, prioritising work over other commitments as appropriate.

Attention to detail

Extent to which candidate is thorough in accomplishing a task, pays close attention to detail, and takes a systematic unhurried approach.

Ability to check and not make assumptions

The extent to which the candidate, before taking action, makes sure that it is a sensible action to take and that the required information is available, accurate and used correctly.

Proactive

The extent to which the candidate takes initiative when reporting or dealing with issues, anticipates problems and takes appropriate actions, accepts responsibility for own actions and does not over-rely on others.

Tenacious

Determination and perseverance to get the job done in time and do it properly, even in the face of difficulties.

Tolerance of working alone

Sociability

The extent to which the candidate interacts effectively with others without being overly keen for other peoples' company and/or approval.

Sensation seeking

The extent to which the candidate maintains his or her performance, interest and attention in monotonous situations without seeking excitement or acting/making changes on impulse.

Need for external stimulation (extraversion)

The extent to which the candidate maintains his or her performance, interest and attention in monotonous situations without the need for stimulation (from social interaction or other external stimuli).

Now that we understand how you will be assessed during the train driver selection process, let's now take a look at the tests that are currently being used.

TRAIN DRIVER OBSERVATIONAL ABILITY TEST (OAT)

Attention is crucial for any train driver and therefore is heavily assessed in the train driver selection process. One of the methods of this assessment is achieved by practising our Observational Ability Test.

The Observational Ability Test (OAT) provides a measure that is sensitive to sustained attention, selective attention and attentional switching. Your attention will be tested by engaging with familiar everyday items such as telephone directories and looking for key symbols, this part of the test is called the telephone search.

Another subtest includes searching through a telephone directory (like the test mentioned above), but this time simultaneously counting a run of tones played in the background.

The Observational Ability Test is used to determine different patterns of attentional skills and provides a broad-based indication of the theoretical aspects of attention.

Below are 5 examples of a practice test for you to use in order to practice the necessary skills. We have provided you with these practice questions in order to demonstrate qualities and skills are needed in regards to selective attention and key eye for detail.

Question 1

Find all the repair shops with the area code (01128).

Key

[Hospitals icon] Hospitals	[Airports icon] Airports	[Libraries icon] Libraries
[Shops icon] Shops	[Post offices icon] Post offices	[Cinemas icon] Cinemas
[Fire stations icon] Fire stations	[Cafes icon] Cafes	[Bars icon] Bars
[Repair shops icon] Repair shops	[Tailors icon] Tailors	[Shipping companies icon] Shipping companies
[Plumbers icon] Plumbers		

(01608) 241733 [Hospitals] (01305) 351771 [Airports] (01019) 506963 [Shipping] (01608) 674304 [Fire stations] (01305) 730635 [Bars]

(01538) 625285 [Cafes] (01305) 184228 [Fire stations] (01947) 097165 [Cafes] (01128) 396997 [Tailors] (01105) 860435 [Post offices]

(01128) 110909 [Hospitals] (01650) 871346 [Cinemas] (01191) 111525 [Libraries] (01191) 908304 [Hospitals] (01191) 931142 [Tailors]

(01019) 328522 [Tailors] (01608) 010646 [Post offices] (01608) 023819 [Shops] (01105) 519004 [Fire stations] (01128) 346547 [Libraries]

(01105) 500407 [Post offices] (01203) 449968 [Cinemas] (01608) 286053 [Cafes] (01538) 375319 [Cinemas] (01608) 490966 [Shipping]

(01019) 768673 [Cafes] (01128) 007168 [Tailors] (01947) 009104 [Post offices] (01538) 792897 [Cafes] (01947) 882726 [Shipping]

(01305) 011033 [Shops] (01650) 673414 [Cafes] (01191) 359774 [Shipping] (01191) 243615 [Repair shops] (01947) 930095 [Repair shops]

(01203) 400487 [Hospitals] (01608) 964074 [Bars] (01650) 367336 [Plumbers] (01305) 010021 [Shipping] (01191) 879307 [Tailors]

(01203) 672028 [Fire stations] (01191) 113252 [Fire stations] (01305) 338119 [Repair shops] (01538) 711377 [Shipping] (01191) 134999 [Bars]

(01650) 625071 [Hospitals] (01947) 477520 [Libraries] (01305) 228251 [Cafes] (01105) 318166 [Cafes] (01203) 653217 [Cinemas]

(01203) 049551 [Shipping] (01191) 233223 [Tailors] (01608) 208409 [Bars] (01538) 315977 [Shops] (01947) 776939 [Repair shops]

(01019) 293695 [Post offices] (01019) 860292 [Cafes] (01305) 582278 [Cinemas] (01191) 922067 [Shops] (01191) 904937 [Fire stations]

(01538) 080797 [Cafes] (01191) 310154 [Post offices] (01305) 838437 [Tailors] (01203) 135347 [Libraries] (01305) 251141 [Bars]

(01019) 485488 [Fire stations] (01203) 396319 [Post offices] (01608) 739941 [Airports] (01305) 855098 [Shops] (01105) 071495 [Shops]

(01105) 366659 [Tailors] (01650) 179091 [Tailors] (01019) 572479 [Hospitals] (01608) 366136 [Airports] (01947) 362594 [Bars]

(01105) 454075 [Shipping] (01105) 649363 [Airports] (01191) 247427 [Cafes] (01203) 459880 [Shops] (01305) 078467 [Airports]

(01191) 582527 [Cinemas] (01650) 422690 [Bars] (01650) 862336 [Cafes] (01947) 511882 [Hospitals] (01105) 285421 [Airports]

(01947) 455161 [Cafes] (01305) 344378 [Cinemas] (01191) 094136 [Shipping] (01191) 212965 [Fire stations] (01128) 512443 [Hospitals]

(01650) 861302 [Shipping] (01191) 648494 [Airports] (01650) 033695 [Shops] (01105) 304541 [Libraries] (01608) 138966 [Shipping]

(01128) 472506 [Cafes] (01105) 732401 [Bars] (01947) 755295 [Plumbers] (01650) 524014 [Hospitals] (01191) 127834 [Post offices]

(01019) 527238 [Bars] (01305) 125205 [Post offices] (01203) 009358 [Cinemas] (01203) 307558 [Shops] (01191) 230787 [Cinemas]

(01947) 003227 [Tailors] (01105) 960871 [Shops] (01650) 416290 [Libraries] (01650) 924069 [Repair shops] (01128) 056472 [Repair shops]

(01128) 899563 [Shops] (01105) 510661 [Repair shops] (01650) 564378 [Shops] (01947) 135082 [Fire stations] (01191) 402008 [Post offices]

(01191) 526065 [Cafes] (01608) 244746 [Repair shops] (01608) 060159 [Fire stations] (01128) 492231 [Fire stations] (01128) 638348 [Cafes]

Answer ________________

Question 2

Find all the plumbers with the area code (01246).

Key

🏥 Hospitals   ✈️ Airports   📖 Libraries   🛒 Shops   ✉️ Post offices   🎬 Cinemas

🔥 Fire stations   🍴 Cafes   🍸 Bars   🔨 Repair shops   ✂️ Tailors

🚚 Shipping companies   🔧 Plumbers

(01497) 849189 🏥 (01762) 890217 ✉️ (01122) 110525 📖 (01508) 829804 🔥 (01497) 473515 🍴
(01475) 548756 ✈️ (01917) 945288 🔥 (01246) 581782 🔨 (01508) 889953 🍸 (01951) 892007 🍸
(01972) 311189 🎬 (01951) 532280 🔧 (01508) 120905 🎬 (01257) 769388 🍴 (01917) 235406 ✉️
(01246) 560703 🔧 (01475) 247223 🏥 (01972) 458873 🛒 (01972) 344506 ✉️ (01951) 668522 🚚
(01257) 544161 ✈️ (01951) 069950 🚚 (01497) 594397 🛒 (01972) 758940 🔨 (01917) 938445 ✉️
(01246) 342373 ✂️ (01122) 676514 🍸 (01122) 604282 🍴 (01475) 200368 🔧 (01508) 592220 ✉️
(01246) 519155 🍴 (01917) 471627 ✈️ (01951) 491713 📖 (01246) 718998 🍸 (01972) 247678 🔨
(01917) 876378 🍴 (01497) 577847 🔨 (01257) 197755 🚚 (01972) 546172 🍸 (01972) 892019 🍴
(01508) 934166 🍸 (01257) 188735 🔨 (01951) 273298 🔧 (01917) 595464 🔧 (01508) 986853 📖
(01972) 875725 ✉️ (01122) 425304 📖 (01122) 564540 ✉️ (01257) 657991 📖 (01122) 414638 🎬
(01972) 610341 🛒 (01497) 014945 🛒 (01762) 948317 🔧 (01257) 592289 🛒 (01475) 815967 🏥
(01972) 740110 ✈️ (01951) 308068 🏥 (01475) 417235 🍴 (01917) 334413 🍴 (01122) 567819 🔧
(01972) 921437 📖 (01257) 120195 🚚 (01257) 490648 🔧 (01917) 365809 🔥 (01475) 162198 🍸
(01972) 449099 🍴 (01972) 439554 🔥 (01497) 258274 ✈️ (01497) 930221 ✈️ (01246) 745977 📖
(01497) 640222 ✉️ (01246) 352058 🔧 (01122) 419762 🔧 (01246) 929724 📖 (01972) 728000 🚚
(01917) 600825 🎬 (01475) 207453 🍸 (01122) 166790 🔧 (01257) 056368 ✉️ (01475) 752280 🔥
(01246) 803348 🚚 (01762) 096534 🔧 (01762) 767479 🎬 (01508) 597146 🍸 (01257) 264592 ✂️
(01497) 215954 🍴 (01497) 674216 🏥 (01122) 509167 🔥 (01497) 442862 📖 (01917) 543755 ✈️
(01951) 477108 🍸 (01475) 307095 🔧 (01246) 970499 📖 (01475) 018751 🔨 (01508) 028953 ✂️
(01475) 104772 ✉️ (01972) 776135 🍴 (01508) 464929 ✈️ (01257) 411835 🚚 (01951) 777505 📖
(01475) 292579 🛒 (01951) 939947 🔥 (01246) 892152 ✉️ (01972) 401274 ✉️ (01951) 289203 🚚
(01475) 645105 🔥 (01917) 811962 🔧 (01122) 583827 🍴 (01762) 556756 🎬 (01257) 071987 🍴
(01508) 997103 ✂️ (01246) 497806 🔧 (01762) 330222 🍴 (01257) 983973 🔥 (01951) 167478 🔧
(01122) 234081 🍴 (01246) 364527 🎬 (01762) 277927 ✂️ (01972) 680922 ✉️ (01762) 059172 🎬

Answer [ ]

Question 3

Find all of the cafes with the area code (01362).

Key

Hospitals   Airports   Libraries   Shops   Post offices   Cinemas

Fire stations   Cafes   Bars   Repair shops   Tailors

Shipping companies   Plumbers

(01764) 239633 ⚒ (01362) 966687 🚚 (01916) 110316 ⚒ (01865) 833776 ✚ (01362) 512193 🛒
(01294) 430650 ✈ (01294) 686282 ✂ (01362) 235335 🍴 (01879) 848735 🔧 (01865) 111028 🔥
(01916) 415008 ✚ (01777) 774569 🔧 (01916) 842063 🔧 (01294) 399229 ✂ (01879) 188939 📖
(01916) 905506 ⚒ (01610) 791405 🍴 (01294) 842426 🛒 (01359) 150894 🚚 (01879) 936207 ✂
(01411) 192702 ✉ (01865) 032885 ⚒ (01610) 332875 🔥 (01764) 137278 🎞 (01359) 976940 🔧
(01916) 444531 ✂ (01879) 252365 🚚 (01865) 910921 🔥 (01777) 774958 🍴 (01359) 051766 ✂
(01294) 168313 🛒 (01865) 512266 🔥 (01916) 442668 ✚ (01411) 191317 ✂ (01359) 202712 🍸
(01294) 287109 ✚ (01362) 127785 🛒 (01764) 906538 🔧 (01294) 834667 📖 (01916) 128892 🍸
(01916) 750762 🚚 (01916) 104420 ✈ (01865) 435393 🍸 (01610) 777304 ✈ (01879) 102277 🚚
(01879) 807062 ✉ (01764) 549051 🍸 (01610) 532435 ⚒ (01865) 332995 🚚 (01916) 356345 📖
(01294) 337247 🔧 (01777) 938924 🚚 (01610) 570284 🛒 (01879) 351073 🍸 (01362) 162429 🎞
(01777) 067561 ⚒ (01362) 804123 🍸 (01879) 014169 🍴 (01359) 139089 ✂ (01916) 196636 ✉
(01879) 516465 🍴 (01610) 566638 ⚒ (01362) 209496 🔥 (01411) 604498 🔥 (01294) 662097 🎞
(01777) 895864 🚚 (01764) 729988 🔧 (01359) 465583 🔥 (01359) 655875 🔥 (01294) 040042 ⚒
(01610) 813548 🛒 (01879) 767593 🔧 (01777) 795881 🔧 (01359) 818049 📖 (01916) 647126 ✈
(01916) 678666 🚚 (01879) 441343 🍸 (01879) 112354 🔥 (01411) 061459 🔥 (01879) 190318 ✂
(01865) 794846 📖 (01411) 163554 ⚒ (01764) 443582 🍴 (01362) 975727 🔧 (01879) 728689 🛒
(01865) 345141 🛒 (01362) 630057 ✉ (01865) 091197 🔥 (01359) 782471 ✉ (01764) 480505 🚚
(01764) 127061 🛒 (01294) 001747 🍴 (01879) 388711 ✈ (01610) 303416 🎞 (01916) 674985 📖
(01610) 255452 🔧 (01777) 177421 🛒 (01764) 460730 🛒 (01359) 691744 🔧 (01764) 362377 🍴
(01777) 677520 ✈ (01865) 666623 📖 (01362) 658169 🔥 (01764) 218323 🚚 (01865) 921346 ✉
(01777) 324258 ⚒ (01610) 441014 ✈ (01359) 629910 ✉ (01777) 066769 🔧 (01294) 669730 ✚
(01777) 513478 ✉ (01879) 347560 ✈ (01294) 008964 🛒 (01879) 448008 🍸 (01359) 378665 🍸
(01865) 897540 ⚒ (01359) 191426 🛒 (01777) 437672 🛒 (01865) 772225 🚚 (01879) 198553 🍸

Answer

Question 4

Find all of the shops with the area code (01926).

Key

Hospitals   Airports   Libraries   Shops   Post offices   Cinemas

Fire stations   Cafes   Bars   Repair shops   Tailors

Shipping companies   Plumbers

(01057) 040637 Cafes (01926) 070081 Shops (01142) 398726 Shipping (01926) 137949 Fire station (01535) 416284 Cinema
(01144) 047838 Cinema (01694) 514966 Hospital (01966) 524228 Hospital (01275) 298184 Hospital (01142) 203392 Airport
(01275) 591078 Fire station (01887) 077583 Tailor (01142) 988561 Airport (01144) 488639 Cinema (01057) 519558 Repair shop
(01887) 523139 Shops (01288) 945321 Plumber (01694) 360404 Bars (01926) 574227 Library (01535) 273467 Cafes
(01887) 741592 Hospital (01694) 932906 Cafes (01535) 605255 Tailor (01275) 265805 Library (01142) 575450 Shipping
(01142) 236216 Repair shop (01142) 752906 Bars (01887) 145924 Cafes (01144) 909828 Shops (01144) 616119 Tailor
(01535) 562136 Cafes (01887) 400497 Fire station (01887) 202381 Bars (01535) 432360 Library (01926) 829119 Airport
(01966) 715604 Hospital (01887) 436538 Plumber (01275) 959842 Fire station (01535) 483979 Plumber (01535) 289803 Plumber
(01142) 052541 Post office (01926) 592728 Tailor (01887) 403319 Post office (01694) 064077 Shops (01142) 463141 Cinema
(01275) 402908 Shipping (01275) 214803 Cafes (01694) 799647 Cinema (01694) 050279 Library (01966) 503123 Library
(01926) 746781 Airport (01926) 843129 Cafes (01275) 305461 Plumber (01144) 717968 Plumber (01275) 691216 Cafes
(01275) 093721 Shops (01288) 139692 Library (01142) 358707 Shops (01142) 870473 Tailor (01144) 670000 Tailor
(01057) 709553 Cafes (01887) 111657 Cinema (01887) 733173 Cafes (01144) 983247 Hospital (01275) 988382 Bars
(01288) 048305 Post office (01057) 554789 Cinema (01535) 807212 Fire station (01142) 551007 Bars (01057) 405393 Plumber
(01926) 853626 Library (01144) 090414 Fire station (01535) 036461 Fire station (01887) 923379 Library (01926) 569660 Shops
(01288) 915875 Tailor (01694) 553815 Library (01142) 063113 Post office (01142) 093486 Shops (01926) 738734 Repair shop
(01535) 689512 Plumber (01275) 824220 Shops (01694) 217898 Airport (01057) 854190 Plumber (01142) 283640 Hospital
(01966) 248473 Cafes (01144) 846544 Cafes (01966) 714938 Repair shop (01057) 873628 Shipping (01887) 163833 Shops
(01926) 775148 Bars (01144) 229688 Library (01144) 979010 Cafes (01887) 743542 Shops (01966) 269200 Library
(01966) 270905 Airport (01887) 690206 Post office (01057) 092772 Bars (01057) 225300 Plumber (01926) 372883 Fire station
(01926) 844866 Post office (01142) 286285 Shipping (01142) 844109 Repair shop (01142) 001946 Cinema (01144) 322642 Fire station
(01694) 483342 Plumber (01288) 800048 Airport (01966) 879737 Hospital (01288) 988672 Airport (01966) 301677 Cinema
(01142) 131436 Shops (01966) 943104 Tailor (01057) 327639 Bars (01926) 370025 Bars (01288) 979304 Repair shop

Answer

Question 5

Find all of the bars with the area code (01087).

Key

[Hospitals] Hospitals	[Airports] Airports	[Libraries] Libraries	[Shops] Shops
[Post offices] Post offices	[Cinemas] Cinemas		
[Fire stations] Fire stations	[Cafes] Cafes	[Bars] Bars	[Repair shops] Repair shops
[Tailors] Tailors			
[Shipping companies] Shipping companies	[Plumbers] Plumbers		

(01983) 531007 [Hospitals] (01659) 023255 [Hospitals] (01862) 691881 [Plumbers] (01964) 077081 [Post offices] (01964) 566236 [Hospitals]

(01735) 806729 [Repair shops] (01659) 999075 [Libraries] (01231) 759170 [Airports] (01087) 414782 [Cinemas] (01852) 092499 [Hospitals]

(01830) 354392 [Plumbers] (01087) 935038 [Bars] (01231) 725875 [Libraries] (01199) 511629 [Hospitals] (01852) 798485 [Cafes]

(01659) 209970 [Fire stations] (01964) 844670 [Cafes] (01087) 497741 [Hospitals] (01852) 898628 [Post offices] (01862) 601403 [Shipping companies]

(01830) 942835 [Shipping companies] (01087) 383029 [Post offices] (01862) 378617 [Fire stations] (01830) 568733 [Libraries] (01964) 155893 [Airports]

(01659) 563124 [Hospitals] (01231) 938749 [Airports] (01983) 461120 [Repair shops] (01983) 920418 [Libraries] (01852) 312354 [Tailors]

(01199) 722632 [Airports] (01199) 318066 [Fire stations] (01199) 086162 [Plumbers] (01964) 042345 [Hospitals] (01964) 423938 [Plumbers]

(01830) 200339 [Fire stations] (01659) 987205 [Post offices] (01735) 642101 [Repair shops] (01735) 277814 [Post offices] (01231) 255231 [Fire stations]

(01659) 569376 [Airports] (01830) 882056 [Libraries] (01231) 200847 [Libraries] (01231) 388472 [Post offices] (01231) 192646 [Airports]

(01735) 090890 [Shipping companies] (01087) 122007 [Tailors] (01199) 069056 [Shipping companies] (01199) 414836 [Libraries] (01964) 136442 [Libraries]

(01964) 469045 [Cinemas] (01852) 916766 [Bars] (01659) 775830 [Cinemas] (01087) 304251 [Tailors] (01231) 033096 [Fire stations]

(01087) 620842 [Libraries] (01659) 462557 [Hospitals] (01735) 691743 [Cinemas] (01735) 172599 [Hospitals] (01087) 017250 [Shops]

(01659) 996026 [Tailors] (01964) 930486 [Shipping companies] (01231) 476818 [Libraries] (01983) 487310 [Bars] (01199) 979485 [Plumbers]

(01852) 393910 [Post offices] (01964) 008423 [Tailors] (01862) 198258 [Bars] (01830) 002058 [Shipping companies] (01862) 567982 [Repair shops]

(01862) 118203 [Plumbers] (01862) 977878 [Hospitals] (01199) 645659 [Plumbers] (01735) 023472 [Bars] (01735) 399317 [Libraries]

(01231) 449324 [Shipping companies] (01862) 684362 [Post offices] (01862) 510565 [Fire stations] (01964) 189181 [Bars] (01862) 029218 [Cinemas]

(01231) 460995 [Airports] (01231) 656368 [Fire stations] (01199) 889256 [Cafes] (01964) 447342 [Fire stations] (01862) 095663 [Plumbers]

(01231) 457499 [Bars] (01862) 981464 [Repair shops] (01087) 382241 [Hospitals] (01852) 508290 [Plumbers] (01830) 401436 [Repair shops]

(01199) 953532 [Libraries] (01852) 415937 [Airports] (01983) 631030 [Post offices] (01087) 840312 [Shops] (01862) 695597 [Plumbers]

(01862) 362717 [Tailors] (01231) 072794 [Post offices] (01852) 429249 [Bars] (01983) 311828 [Plumbers] (01964) 142454 [Airports]

(01830) 629112 [Cinemas] (01830) 012149 [Bars] (01087) 734209 [Tailors] (01087) 317509 [Fire stations] (01199) 922286 [Cafes]

(01087) 237046 [Post offices] (01087) 344774 [Fire stations] (01231) 744794 [Shops] (01830) 532830 [Fire stations] (01087) 180272 [Repair shops]

(01735) 211103 [Libraries] (01659) 863265 [Shipping companies] (01983) 309535 [Hospitals] (01830) 037144 [Libraries] (01830) 798886 [Post offices]

(01862) 064992 [Fire stations] (01087) 610014 [Bars] (01964) 631568 [Cafes] (01862) 183130 [Bars] (01852) 114457 [Plumbers]

Answer ______________

Answers to TEA-OCC

Q1. 1

(01947) 003227 ✂ (01105) 960871 🛒 (01650) 416290 📖 (01650) 924069 ✎ **(01128) 056472 ✎**
(01128) 899563 🛒 (01105) 510661 ✎ (01650) 564378 🛒 (01947) 135082 🔥 (01191) 402008 ✉
(01191) 526065 🍴 (01608) 244746 ✎ (01608) 060159 🔥 (01128) 492231 🔥 (01128) 638348 🍴

Q2. 3

(01475) 548756 ✈ (01917) 945288 🔥 (01246) 581782 ✎ (01508) 889953 ⚒ (01951) 892007 ⚒
(01972) 311189 🎴 (01951) 532280 🔧 (01508) 120905 🎴 (01257) 769388 🍴 (01917) 235406 ✉
(01246) 560703 🔧 (01475) 247223 ⚕ (01972) 458873 🛒 (01972) 344506 ✉ (01951) 668522 🚚

(01972) 449099 🍴 (01972) 439554 🔥 (01497) 258274 ✈ (01497) 930221 ✈ (01246) 745977 📖
(01497) 640222 ✉ **(01246) 352058 🔧** (01122) 419762 ✎ (01246) 929724 📖 (01972) 728000 🚚

(01508) 997103 ✂ **(01246) 497806 🔧** (01762) 330222 🍴 (01257) 983973 🔥 (01951) 167478 ✎
(01122) 234081 🍴 (01246) 364527 🎴 (01762) 277927 ✂ (01972) 680922 ✉ (01762) 059172 🎴

Q3. 1

(01764) 239633 ✎ (01362) 966687 🚚 (01916) 110316 ✎ (01865) 833776 ⚕ (01362) 512193 🛒
(01294) 430650 ✈ (01294) 686282 ✂ **(01362) 235335 🍴** (01879) 848735 🔧 (01865) 111028 🔥
(01916) 415008 ⚕ (01777) 774569 🔧 (01916) 842063 🔧 (01294) 399229 ✂ (01879) 188939 📖

Q4. 2

(01057) 040637 🍴 **(01926) 070081 🛒** (01142) 398726 🚚 (01926) 137949 🔥 (01535) 416284 🎴
(01144) 047838 🎴 (01694) 514966 ⚕ (01966) 524228 ⚕ (01275) 298184 ⚕ (01142) 203392 ✈
(01275) 591078 🔥 (01887) 077583 ✂ (01142) 988561 ✈ (01144) 488639 🎴 (01057) 519558 ✎

(01926) 853626 📖 (01144) 090414 🔥 (01535) 036461 🔥 (01887) 923379 📖 **(01926) 569660 🛒**
(01288) 915875 ✂ (01694) 553815 📖 (01142) 063113 ✉ (01142) 093486 🛒 (01926) 738734 ✎
(01535) 689512 🔧 (01275) 824220 🛒 (01694) 217898 ✈ (01057) 854190 🔧 (01142) 283640 ⚕

Q5. 2

(01983) 531007 🧰 (01659) 023255 🧰 (01862) 691881 🔧 (01964) 077081 ✉ (01964) 566236 🧰
(01735) 806729 🔧 (01659) 999075 📖 (01231) 759170 ✈ (01087) 414782 📧 (01852) 092499 🧰
(01830) 354392 🔧 (01087) 935038 ⌛ (01231) 725875 📖 (01199) 511629 🧰 (01852) 798485 🍴

(01087) 237046 ✉ (01087) 344774 🔥 (01231) 744794 🛒 (01830) 532830 🔥 (01087) 180272 🔧
(01735) 211103 📖 (01659) 863265 🐾 (01983) 309535 🧰 (01830) 037144 📖 (01830) 798886 ✉
(01862) 064992 🔥 (01087) 610014 ⌛ (01964) 631568 🍴 (01862) 183130 ⌛ (01852) 114457 🔧

DOTS CONCENTRATION TEST

The Dots Concentration Test is a Train Driver test that can be used to improve your concentration skills. It is designed to measure a train driver's ability to maintain concentration and accuracy for a long period of time. As a train driver you have the responsibility for the safety of yourself, the crew members and the passengers, which is of the upmost importance. Due to this, the test is often used to eliminate many candidates from the selection process early on.

The test is a psychometric test used to access your ability to maintain diligence, speed, accuracy and concentration whilst looking for a cluster of 4 dots. If you chose groups of dots other than that of groups of four, you will lose marks. Consistency is key. Scoring well on the test will show the recruiters that you have the qualities needed to become a safe train driver.

VIGILANCE TEST

For this practice test, we will provide you with visual tests to assist you in your preparation for your Train Driver assessment. It is essentially a test to measure your attention to detail which is a vital skill for any aspiring Train Driver.

ASSESSING AND CHECKING INFORMATION TESTS

Train Drivers will have to undertake tests that focus on their ability and regulations. The Assessing and Checking Information Tests can be used as a practice tool only, and is in no way a portrayal of the real test. Primarily, we have designed these tests to help you through your preparation and enable you to practice the key skills and attributes required for the role of a Train Driver. Whilst we have provided you with sample questions regarding faults, errors and checks, you need to be aware that these tests are not an exact replica of the real test. Understand that we have provided these sample tests to help you better your trainability and thus improve your overall chances of success.

The test will examine your ability to maintain and recall job related knowledge. Testing your memory is a crucial stage to the application process of becoming a Train Driver. Employers want to see strong candidates who show high levels of memory and concentration skills. A good performance on this type of test can often be linked with train driving performance and therefore the test has been identified to have a significant impact on the selection process of train drivers in the UK. The test will also examine your ability to learn new concepts that represent those often encountered in driver training.

Ultimately, this test can be used as a way of practising analysis for finding faults in a set of data; a skill that will be essential to the role of a successful Train Driver.

TRAIN DRIVER AWARENESS AND RECOGNITION TEST (AART)

This test has been designed to test your observational ability by quickly presenting pictures of traffic conditions on screen. We have created a test whereby you will be presented a picture that should be looked at no more than 1 second, and after seeing each picture, you will be asked to identify what the picture contained from a list of 5 options. The process then repeats.

This test will assess your speed of perception, visual observation and ability to maintain an overview of that observation.

During the test you will see various pictures for a short period of time. You will then be asked to answer questions about what you saw. Take a look at the following sample image which will appear for a brief second during the test:

QUESTION: What did you see in the picture?

1. A car

2. A van

3. A pedestrian

4. A cyclist

5. A truck

You will then need to select which objects that you saw in the picture. In this case the answers you would need to select would be as follows:

2. A car

4. A cyclist

On the next 5 pages are examples of Awareness and Recognition Tests in order for you to try out. Look at the picture and remember what you see. You will be asked questions based on the image, so it is crucial you look at the image closely.

Look at the image for no longer than 1 second. Cover up the image and answer the following question in association with that image. Answers will be provided at the end of the section. You can select more than one answer per question.

Question 1

What did you see in the picture?

A = Three cars

B = Four cars

C = Road junction

D = A motorcycle

E = A cyclist

Answer

Question 2

What did you see in the picture?

A = Cycle lane

B = Bus lane

C = Road signs

D = Traffic lights

E = pedestrians

Answer

Question 3

What did you see in the picture?

A = Single yellow lines

B = Pedestrians

C = A cyclist

D = A roundabout

E = A lorry

Answer

Question 4

What did you see in the picture?

A = A pedestrian

B = A Shell Garage

C = A cyclist

D = Parked cars

E = A Ford Garage

Answer

Question 5

What did you see in the picture?

A = A motorcyclist

B = Two bikes

C = A rubbish bin

D = A cyclist

E = A pedestrian

Answer

ANSWERS TO AWARENESS AND RECOGNITION TEST (AART)

Q1. B and C

Q2. A, C, D and E

Q3. D

Q4. D

Q5. B, C and D

REACTION TIMES

The trainee train driver selection process assesses your ability to react fast and accurately through the Reaction Times Test. It is utilised to measure how well you can concentrate intently whilst you react to timed events presented on screen. It is a computerised test in which you are first assessed on your own pace and then you are assessed at the computer's pace. It will test your ability to focus on the task at hand by having the ability to continue the test fully focussed even if you realise you've made a mistake.

CO-ORDINATION TEST FOR TRAIN DRIVERS

The purpose of the test is to reliably assess your eye-to-hand and hand-to-hand co-ordination. The task will involve you controlling a red dot which you have to guide along a given track. This will be done by using either two control knobs or joysticks. The track in which you need to guide the red dot along consists of three sections that will make various demands on your ability to co-ordinate your left and right hand. This is the greatest difficulty of the task; being able to co-ordinate both hands in the need to make the correct visual assessment. Anticipation of direction also plays a big part.

WRITTEN COMMUNICATION TEST (WCT)

The WCT is designed by How2become to test the train driver's communicational skills with other people in a written form. It will test how clear and informative your written communication is. Written communication skills are indeed important to the train driver role and it does coincide with safety but the vast majority of a train driver's written tasks are completed whilst not in control of a train i.e. during a work break or completed at the depot. Therefore, written communication is not time critical and a lack of communication skill will not result in a train driving error.

RELEVANT BEHAVIOURAL/PERSONALITY ATTRIBUTES SITUATIONAL JUDGEMENT EXERCISE (SJE)

The situational judgement exercise (SJE) is a multiple choice test that presents a range of different situations that you encounter in an actual train driver role. The exercise will ask you to select the most appropriate behavioural responses to those situations, or the behavioural responses that you yourself would chose. There are normally between 4-5 possible actions to choose from and it will be your task to distinguish which is the most effective course of action to take (i.e. the action you would make yourself if the situation were real).

Train operating companies' use SJEs as part of the selection process as it provides strong evidence for situation-based measures in predicating your job performance. It will be used to identify the potential strong performers within the selection process.

You can try out Situational Judgement Tests in the following guide, *Train Driver Situational Judgement Tests*:

Below are 5 examples of Situational Judgement Tests (SJTs) in order for you to try out. For the following questions in this chapter, write the letter of the sentence that matches the course of action you would most likely take.

QUESTION 1

Whilst serving as a fully qualified Train Driver, you become concerned that the standard of continuous professional development training has been dropping over the last few months. What do you do?

ANSWER OPTIONS

A. Provide feedback to your manager on how you believe the training could be improved in order to improve safety.

B. At the end of the day, any issues surrounding training standards are down to the line manager. It is up to them to sort it out, and if they can't recognise there's an issue, then that's not your problem.

C. Apply for promotion so that you can be in charge of training and therefore improve the standards.

D. Discuss your concerns with the other Train Drivers first. If they agree with you, then inform your manager.

ANSWER OPTIONS IN ORDER OF PRIORITY

1st	
2nd	
3rd	
4th	

QUESTION 2

Whilst preparing the train at Euston Station for your regular weekly journey to Glasgow Central, the train guard approaches you in the cab to inform you that there are problems with the train's communication system. He tells you that no announcements will be able to be made to passengers during the journey because of the fault. What would you do?

ANSWER OPTIONS

A. Thank the guard for the information and carry on with your duties preparing the train.

B. Thank the guard for the information before informing the station manager about the defect. You will not be able to drive this train, as you will not be able to make any safety-related announcements to the passengers during the journey, if the need arises. The safety of the passengers and the other crew members is paramount.

C. Thank the guard for the information before heading off to see if you can fix the problem yourself. You are a Train Driver, and as such, you are a highly practical and adaptable person. You are confident you can fix the problem before the train departs the platform. If you can't fix it, you will then inform your station manager of the problem.

D. Thank the guard for the information before informing him that you will contact him directly if any important announcements need to be made during the journey. If there are any announcements that need to be made, the guard can then walk up and down the train carriages informing the passengers. The most important thing is to ensure the passengers receive all important messages.

ANSWER OPTIONS IN ORDER OF PRIORITY

1st	
2nd	
3rd	
4th	

QUESTION 3

You wake up one morning and start to get ready for work. You are scheduled to drive the 10:05 train from Paddington to Reading. All of a sudden you start to feel ill and uneasy. Whilst driving to work your eyes become slightly blurred, but you are just about OK to drive. When you arrive at work you still feel slightly ill. What would you do?

ANSWER OPTIONS

A. Immediately inform your manager about how you feel and that you think you're OK to drive the train.

B. Battle through it. You are a Train Driver and you have a responsibility to drive the train on time, so that the passengers get to their destination.

C. Take headache tablets and monitor the situation. If you feel bad during the journey from Paddington to Reading, then you will consider booking sick once you reach Reading.

D. Immediately book sick so that a replacement driver can be found, before informing your line manager.

ANSWER OPTIONS IN ORDER OF PRIORITY

1st	
2nd	
3rd	
4th	

QUESTION 4

You are a fully qualified and experienced Train Driver. You are sat around the station canteen mess table having a cup of tea with some other Train Drivers. There's a bit of banter going on and one of the more senior Train Drivers is making fun out of a newly qualified Train Driver. The new Train Driver appears to be slightly upset by the comments but doesn't say anything. What would you do?

ANSWER OPTIONS

A. Intervene and say that you do not think the comments are appropriate. Even though the driver is new, that doesn't mean he should be the centre of any harmful jokes or comments.

B. Banter is all part of railway working life. Train Drivers work in lonely and highly pressurised situations and a bit of canteen banter is good for morale. If the new Train Driver is going to fit in, then he will need to get used to it.

C. Leave the canteen and go and inform your station manager of the comments.

D. You would probably join in with the banter.

ANSWER OPTIONS IN ORDER OF PRIORITY

1st	
2nd	
3rd	
4th	

QUESTION 5

You are driving a train with Overhead Line Equipment (OLE) between Slough and Paddington when you suddenly notice in the distance a metal scaffolding bar hanging from the catenary wire. The metal scaffolding bar looks like it will hit your cab screen window. What would you do?

ANSWER OPTIONS

A. Immediately bring the train to a safe stop and report the incident to the signaller and the Electrical Control Operator (ECO). You would then follow your training and inform the guard and passengers that you had to stop due to an incident and that you would update them as soon as you had any further information. Under no circumstances would you leave the cab, for safety reasons.

B. Immediately bring the train to a safe stop and report the incident to the signaller and the Electrical Control Operator (ECO). You would then follow your training and inform the guard and passengers that you had to stop due to an incident and that you would update them as soon as you had any further information. You would then leave the cab to assess the situation and see if it was possible to remove the metal bar yourself.

C. Immediately bring the train to a safe stop. You would then leave the cab to assess the situation and see if it was possible to remove the metal bar yourself. This way, you would be able to get moving again quickly to ensure all passengers arrived at their destination on time.

D. As you understand, responsibility for safety rests with the train guard, and as such, he/she would be responsible for dealing with this type of incident. You would stop the train and then would ask them what they intended to do about it.

ANSWER OPTIONS IN ORDER OF PRIORITY

1st	
2nd	
3rd	
4th	

Answers to Situational Judgement Tests (SJTs)

SUGGESTED ANSWERS IN ORDER OF PRIORITY

QUESTION 1

A - This answer is best since you are addressing the problem that you've noticed by constructively giving feedback.

D - This answer is better, but not perfect. You are addressing the problem in some way, but would be better off addressing it directly rather than discussing it with others first.

C - This is not an effective solution because it does not address the problem directly.

B - This answer option completely ignores what could be a serious problem. Therefore, it's the least effective course of action.

QUESTION 2

B - This answer option prioritises safety of personnel and passengers. Therefore, it is the best course of action.

D - This answer option prioritises safety of personnel and passengers, but is not as efficient as answer option B.

C - This answer does not prioritise the safety or personnel and passengers. Therefore, it is not an optimal course of action.

A - This is the worst answer since it completely disregards a fault in the train.

QUESTION 3

D - This answer option promotes the safety of yourself, personnel, and passengers, by recognising that you might not be at peak performance.

A - This isn't the best answer option, but at the very least you are informing your manager of the situation.

C - This is inefficient since booking sick halfway through a shift could cause major disruption.

B - This is the least effective answer since you are ignoring your illness, not informing anyone, and potentially putting passengers at risk.

QUESTION 4

A - This is the best course of action because you are taking steps to help prevent a more toxic environment among staff.

C - This is an effective course of action, but it would be better if you were to handle this yourself. An established member of the team calling out inappropriate behaviour could go a long way to solving the problem.

B - This is not an effective solution because while you are not getting involved in the 'banter', you aren't preventing it either.

D - This is the worst course of action because you are contributing to a toxic enviornment in the workplace.

QUESTION 5

A - This answer option places safety as the top priority, resulting in a safer environment for passengers and personnel. Therefore, it is the best course of action.

D - This is a safe course of action, but is less proactive than answer option A.

B - This is not an effective course of action since, as a train driver, you are not to leave the cab. This is for your own safety, as well as the safety of passengers.

C - This is the least safe course of action and gives the impression that you will not ensure that the problem is suitably resolved before moving again.

Multimodal Interview (MMI)

The Multi-Modal Interview is developed to work in conjunction with the Situational Judgement Exercise (SJE). Its purpose is to access your verbal communication and the behavioural selection criteria. You will be assessed on your ability to be able to communicate work related information in a clear and accurate manner. Once again, safety is one of the key elements that the recruiters will assess you on. So being able to verbally communicate safety-critical information is a must.

ASSESSING INFORMATION TEST

The Assessing Information Test is not the same as any real train driver test but it is good preparation. The first part of the test has been designed to assess your memory skills. Train Drivers need to be able to show high levels of memory and concentration skills. You will be required to absorb lots of information at one time, and so it is important that you practice these skills in order to fully prepare for your real assessment. Try to complete this practice test in 7 minutes, in which you should aim to answer all 14 multiple-choice questions that relate to the information you have just read.

Please note, this test is simply to practice your skills and qualities, and is by no means a depiction of the actual test itself.

TEST 1, SECTION 1

You have 4 minutes only to read the following passage and take notes before answering the questions on the following page.

Train tracks are made up of three main components: the metal rails, the sleepers that sit firmly underneath the rails and the ballast – the crushed rock segments that form a bed for the tracks to lie in.

A tamping machine, or a ballast tamper, is a machine that can be used for raising, straightening and altering the tracks. It is a machine that compresses the track ballast (the crushed segments of rock) under the rail track in order to make it more durable. Originally, this work was done manually, whereby labourers would use beaters as a way of pressing down the rock. Despite being faster, more accurate, efficient and less intensive, tamping machines are essential for the production and usage of creating stable tracks from concrete, typically weighing 250 kg.

For train tracks to work efficiently, the alignment of the tracks must be seamless. The sleepers that sit firmly underneath the rails, must also sit firmly in the crushed rock. If a track has been used for many years, or changes to the track have been made, the alignment of all three components needs to be altered in order to remain stable and effective. The gaps in the underlying rock bed need to be filled so that the sleepers do not move as a train passes along. This allows the train to run smoothly and ultimately reduces noise, vibrations and more importantly, any hazards.

Tamping machines can be used to fix these gaps by placing them on the track. It then conducts vibrations with hydraulic 'fingers' to remove all the gaps in the ballast, and align the track up. These machines are very noisy and often cause disruption. Not only are the machines themselves noisy, but they also trigger track alarms which act as a warning sign for workers of approaching trains.

Most tamping jobs are conducted during the night, in order to avoid disrupting train services. By conducting this job at night, therefore can affect nearby neighbourhoods for one or two nights. Due to tamping machines being a part of regular maintenance work, Train Operating Companies are often unable to notify neighbourhoods that may be disturbed.

Question 1

Name **one** component that makes up the train tracks. Three **possible** answers.

Answer

Question 2

What is the name of the machine that is used in order to compress the crushed rock underneath the rail tracks?

A. Tampering machine

B. Tangent machine

C. Hydraulics machine

D. Tamping machine

E. Tramping machine

Answer

Question 3

What is the typical weight of the concrete used in the production process?

A. 150 kg

B. 200 kg

C. 250 kg

D. 300 kg

E. 350 kg

Answer

Question 4

The machine compresses the rock underneath the tracks, in order to make the tracks more….

A. Flexible

B. Diverse

C. Tangent

D. Resistant

E. Durable

Answer

Question 5

Whereabouts are the sleepers positioned?

A. Underneath the ballast

B. To the side of the ballast

C. Underneath the rails

D. To the side of the rails

E. On top of the rails

Answer

Question 6

What is the main job of the tamping machine?

A. To compress the rock

B. To align the rails

C. To generate vibrations

D. To assist the trains mobility

E. To fix any gaps and voids in the track alignment

Answer

Question 7

What does these machines often trigger?

A. Signals

B. Mobility

C. Light

D. Alarms

E. Cannot be determined

Answer

Question 8

The alarms have been triggered. What is the reason for the alarm?

A. The sound of the job completed

B. To warn workers of an approaching train

C. The sound of the job beginning

D. The warn neighbours of work in progress

E. Cannot be determined

Answer

Question 9

What time, day or night, are tamping machines often used?

Answer

Question 10

What does this machinery conduct? **Two** answers needed.

A. Heat

B. Vibrations

C. Light

D. Noise

E. Mobility

Answer

Question 11

How was this job done originally?

Answer

Question 12

How often are the people living close by, notified of work with these machines?

A. Very often

B. Often

C. Neutral

D. Hardly ever

E. Never

Answer

Question 13

What is another name for the type of machine used?

Answer

Question 14

Why might the train track have gaps in the ballast?

A. Animals digging holes underneath the train tracks

B. The rock has eroded

C. The tracks have been used for many years

D. The infrastructure was not made correctly

E. Cannot be determined

Answer

ANSWERS TO TEST 1, SECTION 1

1. Metal rails, the sleepers, or the ballast – you are only asked for one answer, and so any of these answers will be correct.

2. D = tamping machine

3. C = 250 kg

4. E = durable

5. C = underneath the rails

6. E = to fix any gap or void in the track alignment

7. D = alarms

8. B = to warn workers of an approaching train

9. Night

10. B = vibrations and D = noise

11. Manually, with the use of beaters

12. D = hardly ever

13. Ballast tamper

14. C = has been used for many years

TEST 1, SECTION 2

You have 4 minutes only to read the following passage and take notes before answering the questions on the following page.

Freight trains are primarily used to transport cargo and goods, as opposed to transporting passengers. The railway network in Great Britain has been used to transport goods of various types and in various contingencies since the early 19th century. Whilst good traffic in the UK is considerably lower than other countries, it continues to be used, and continues to grow.

Rail freight has become extremely vital in regards to Britain's economic success. It is argued that using rail freight has contributed to over £800 million to the economy. Not only that, but it has also reduced congestion and carbon emissions, and therefore making this use of transportation more environmentally friendly.

Whether it is transporting raw materials for manufacturing purposes, fuels for electrical generations or consumer goods, businesses in the UK rely on freight trains to transport the cargo in an environmentally friendly and efficient way.

The UK has become more reliant on the use of rail freight which provides a faster, safer, greener and efficient way of transporting loads of cargo. It has been said that rail freight is expected to grow in demand by 30% in the next decade. This is equivalent to 240 additional freight trains per day.

In order to maintain and uphold this level of continual growth and demand for freight trains, train operating companies will work in partnership with the government to move cargo transports off of the road, and improve the quality of life by substantially reducing carbon emissions.

It is fact that, on average, a gallon of fuel will move a tonne of goods 246 miles on rail, but only 88 miles by road. Also, each freight train that is used, takes 60 HGV lorries off the road, ultimately helping carbon emissions.

The First World War was also known as 'The Railway War'. Thousands of tonnes of supplies and munitions were distributed all over Great Britain, whereby the supplies were dispatched from ports in the South East to be shipped over to France and the Front Line. A number of programmes were instigated in order for railways to meet the huge demands of the wartime. The Common User Agreement, conducted under the Coal Transport Act of 1917, was an example of a programme that ultimately enabled better railway services. Over 100 train operating companies collaborated on these programmes and worked together in aid of national interest.

Question 1

What do freight trains carry?

A. Passengers

B. Cargo

C. Passengers and cargo

D. Cannot be determined

Answer

Question 2

How long has freight trains been in use in Great Britain?

A. Early 17th century

B. Late 17th century

C. Early 18th century

D. Late 18th century

E. Early 19th century

Answer

Question 3

The use of rail freight for Britain has been extremely vital in regards to…

A. Government success

B. Train Operating Companies becoming more popular

C. Economical success

D. Transport safety

E. Cannot be determined

Answer

Question 4

On average, how much has rail freight contributed to the economy?

A.   £600 million

B.   £800 million

C.   £300 million

D.   £500 million

E.   £900 million

Answer

Question 5

Which two of the following answers can be concluded from rail freight being more environmentally friendly? **Two** answers required.

A.   Reduces carbon emissions

B.   Reduces the use of HGV's

C.   Reduces cost

D.   Reduces congestion

E.   Reduces numerous transportation methods

Answer

Question 6

On average, a gallon of fuel for freight trains can move a tonne of goods how far?

A. 88 miles

B. 100 miles

C. 246 miles

D. 276 miles

E. 44 miles

Answer

Question 7

On average, a gallon of fuel for road usage can move a tonne of goods how far?

A. 246 miles

B. 44 miles

C. 102 miles

D. 88 miles

E. 70 miles

Answer

Question 8

If a freight train is used, how many HGV lorries are taken off the road?

A. 40

B. 60

C. 30

D. 70

E. 80

Answer

Question 9

How much are freight trains expected to grow in demand within one decade?

A. 20%

B. 70%

C. 50%

D. 40%

E. 30%

Answer

Question 10

If freight trains continue to grow at the rate that is expected, how many additional freight trains will be used per day?

A. 210

B. 240

C. 200

D. 180

E. 190

Answer

Question 11

How many train companies collaborated on the programmes that were instigated during the First World War?

A. Over 50

B. Over 60

C. Over 20

D. Over 80

E. Over 100

Answer

Question 12

What was the name of the Act that enabled better railway services during World War I?

A.   Coal Transport Act 1921

B.   Coal Transport Act 1917

C.   Coal Transport Act 1912

D.   Coal Transport Act 1931

E.   Coal Transport Act 1940

Answer

Question 13

What was the First World War also known as?

A.   Britain's War

B.   British Railway War

C.   Front Line War

D.   Railway War

E.   Cannot be determined

Answer

Question 14

Where were the supplies being shipped? **Two** answers required.

A. Germany

B. France

C. Front Line

D. England

E. Cannot be determined

Answer

ANSWERS TO TEST 1, SECTION 2

1.   B = cargo

2.   E = early 19th century

3.   C = economical success

4.   B = £800 million

5.   A = reduces carbon emissions, D = reduces congestion

6.   C = 246 miles

7.   D = 88 miles

8.   B = 60

9.   E = 30%

10. B = 240

11. E = over 100

12. B = Coal Transport Act 1917

13. D = Railway War

14. B = France, C = Front Line

TEST 1, SECTION 3

1. The railway system of Great Britain is one of the oldest in the world. The first steam locomotive was used in Britain, and has become a paramount feature of contemporary society.

2. The arrival of railways has contributed to the dramatic growth in industrialisation during the nineteenth century, and has ultimately had profound impacts on social and economic changes. The railway 'filled a void' that other forms of transport could not. Railways were able to provide an efficient, fast, cost effective and environmentally friendly service that catered for the needs of many people.

3. The history of the railways in Great Britain is remarkable. The incredible changes that have been made over hundreds of years are remarkable, and can only be described as the 'transformation of transport'. In 1804, the first successful steam locomotive ran on wheels, and was used to transport iron across a track of nine miles. Built by Richard Trevithick, the locomotive – 'the Penydarren', was the world's first steam engine to run on rails.

4. Since the early 19th century, railways have continued to develop and are now a strong infrastructure within hundreds of societies. During the First World War, the Government took over and ran the railways until 1921, when private railway companies regained control. During 1921, an Act was passed in Parliament which submerged four railway companies: known simply as the 'British Rail'.

5. During 1940 in the Second World War, the companies effectively worked together to help Britain's war efforts, and in the late 1940s, these railways were nationalised and formed 'British Railways', which was implemented under the Transport Act. In the 50s, society saw a modernised change in regard to railway services. Diesel and electrical trains were introduced, and started to replace the steam locomotive trains.

6. In 1960, the railways were re-organised in order to try and make money. Secondary routes and branch lines closed. As rationalisation took hold, one-third of the pre-1948 networks closed. A giant leap was undertaken in the 70s, and saw the introduction of the 'high-speed diesel-electric' service trains, and by 1990 both main coastal express routes, the East and West Coast Main Lines, had been electrified between central Scotland and London.

7. In 1994, the Channel Tunnel opened and began the service from England to France. This exponential growth in regard to the railway services has considerably changed over the years, and will continue to adapt.

8. In 2011, the number of journeys in Great Britain between 2010 and 2011 reached a record breaking 1.16 billion. As of 2013, British railways are believed to be the second safest in Europe (after Luxembourg), and ultimately deliver a modernised service for both local and both local and national railway routes.

Question 1

Where was the first steam locomotive train used?

A. China

B. Germany

C. France

D. Britain

E. Spain

Answer

Question 2

The railway was able to do what, that other transportation methods could not?

A. Create profits

B. Fill a void

C. Reduce carbon emissions

D. Help Industrialise society

E. Create transport for the middle classes

Answer

Question 3

In what year was the first running steam locomotive train made?

A. 1800

B. 1901

C. 1821

D. 1804

E. 1904

Answer

Question 4

What was the name of the 'mechanical genius' who built the first steam locomotive train?

Answer

Question 5

What was the first steam locomotive train called?

Answer

Question 6

In 1921, an Act in Parliament was implemented and saw the introduction of…

A. 'British Railways'

B. 'Great British Railways'

C. 'The four rails'

D. 'British Rail'

E. Cannot be determined

Answer

Question 7

What was the name of the Act in 1940 that nationalised the four railways?

A. Transportation Act

B. Local Transport Act

C. Transport Act

D. National Transport Act

E. Cannot be determined

Answer

Question 8

What fraction of the pre-1948 services closed in 1960?

A. One half

B. One quarter

C. One fifth

D. One third

E. Three thirds

Answer

Question 9

What type of trains were introduced in 1970?

A. High speed

B. Diesel

C. Electrical

D. Steam

E. All of the above

Answer

Question 10

What opened in 1994?

A. Medway Tunnel

B. Bradway Tunnel

C. The Channel Tunnel

D. Redhill Tunnel

E. Dartford Tunnel

Answer

Question 11

Between 2010 and 2011, the number of journeys reached a record breaking…

A. 1.61 billion

B. 1.16 billion

C. 1.66 billion

D. 1.16 million

E. 6.16 billion

Answer

Question 12

What is the safest railway in Europe?

Answer

Question 13

What service does the Channel Tunnel offer? I.e. England to…?

Answer

Question 14

In what year did private train companies regain control of the railway services?

A. 1984

B. 1904

C. 1911

D. 1927

E. 1921

Answer

ANSWERS TO TEST 1, SECTION 3

1. D = Britain

2. B = fill a void

3. D = 1804

4. Richard Trevithick

5. 'The Penydarren'

6. D = British Rail

7. C = Transport Act

8. D = one third

9. A = high speed

10. C = the Channel Tunnel

11. B = 1.16 billion

12. Luxembourg

13. France

14. E = 1921

Tips for Assessing Information Test

In the build up to the test, read a news passage for a few minutes before getting a friend to ask you questions based on the content.

During the actual test I strongly advise that you take notes. Even though the notes are taken away from you, writing down information helps you to remember it.

CHECKING TESTS

During this practice test, you will be given some indication of the types of faults you could expect if you were to become a train driver. However, this is not to say your assessment will be based on the practice questions we have provided. Whilst this test is not a real depiction of the train driver test you will sit in your selection process, it will provide you with the basic understanding and knowledge regarding what is expected in terms of skills and qualities. This test is primarily designed by us, and is in no way associated with any Train Operating Company, and so you should use these tests as a practice mechanism only and they will not be referred to in the actual test.

Foremost, we have designed this test to allow you to practice your skills regarding faults and analysing information – typical duties that are essential for a position as a Train Driver.

CHECKING TESTS – EXAMPLE

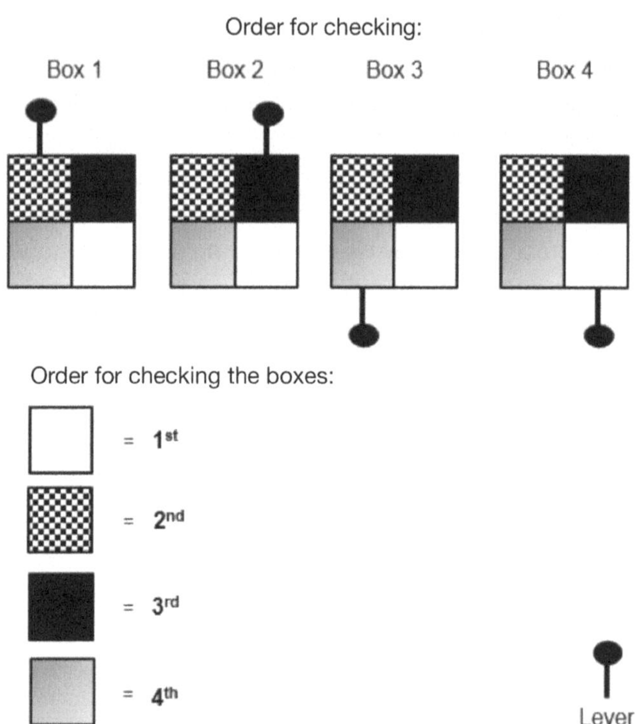

For these types of questions, it is important to take the time and carefully look at the key you are given.

For the above example, you will notice that box 4 would need to be checked first. This is because the lever has been placed in a white box (which indicates that it needs to be made first priority for checking).

The next box that would need to be checked is box 1, then box 2, and then box 3.

Answer: 4123

CHECKING TESTS – TEST SECTION 1

For the following ten questions, write down the order in which the boxes should be checked using the 'Order for Checking' sequence above:

Question 1

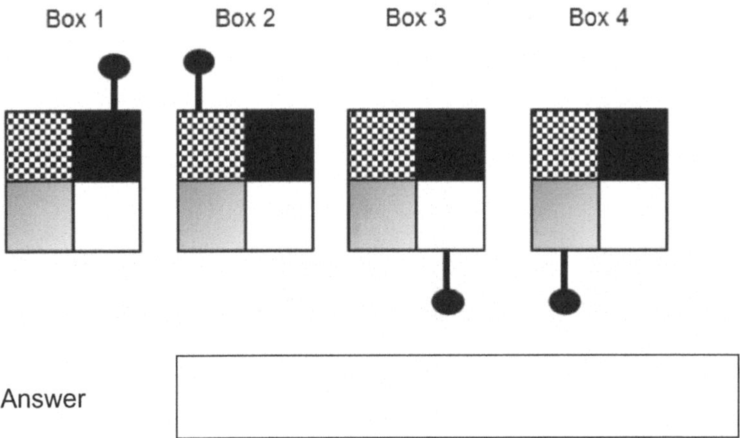

Answer

Question 2

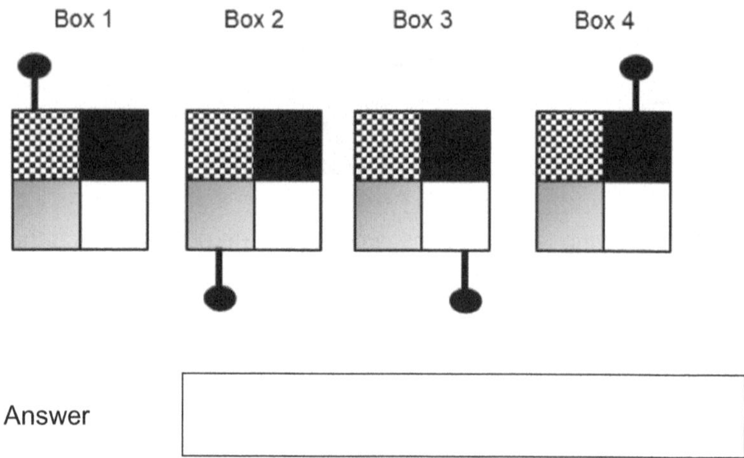

Answer

Question 3

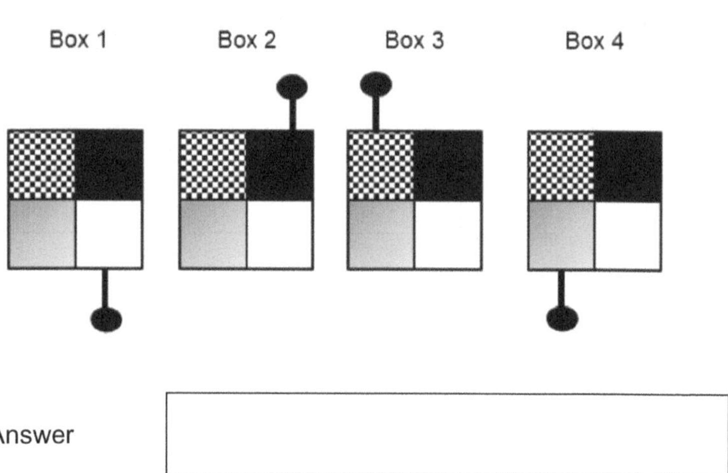

Answer

Question 4

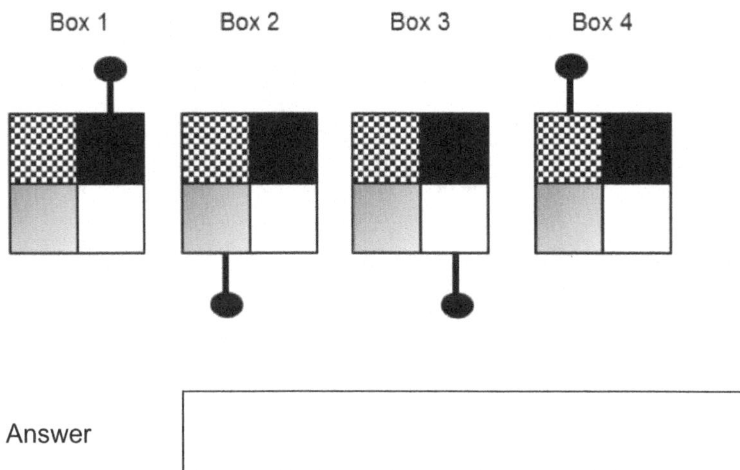

Answer

Question 5

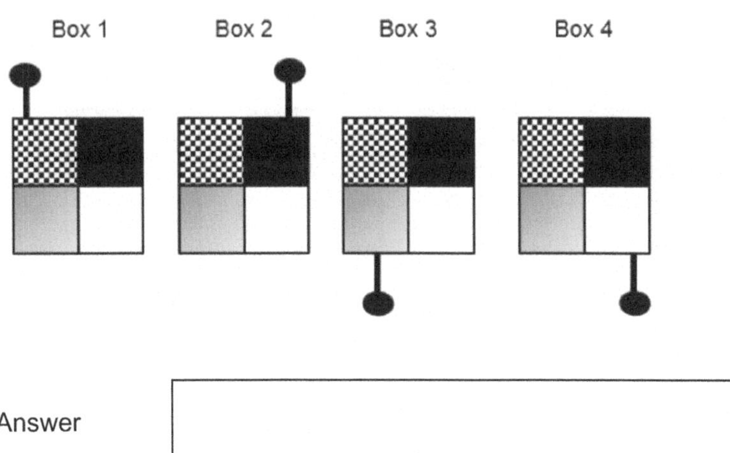

Answer

Question 6

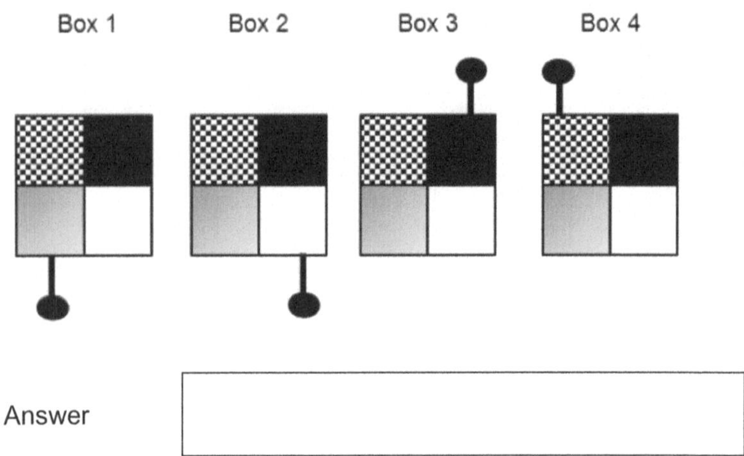

Answer

Question 7

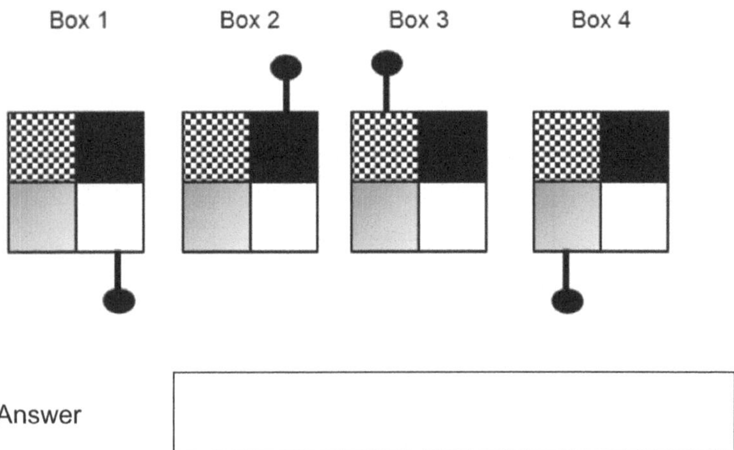

Answer

Question 8

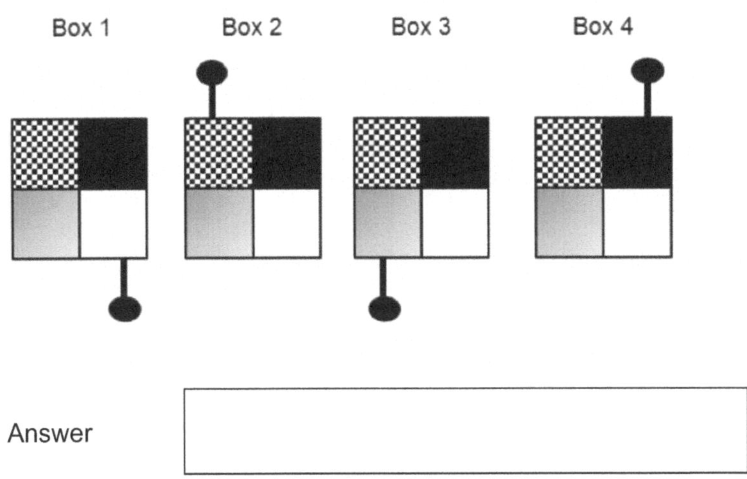

Answer

Question 9

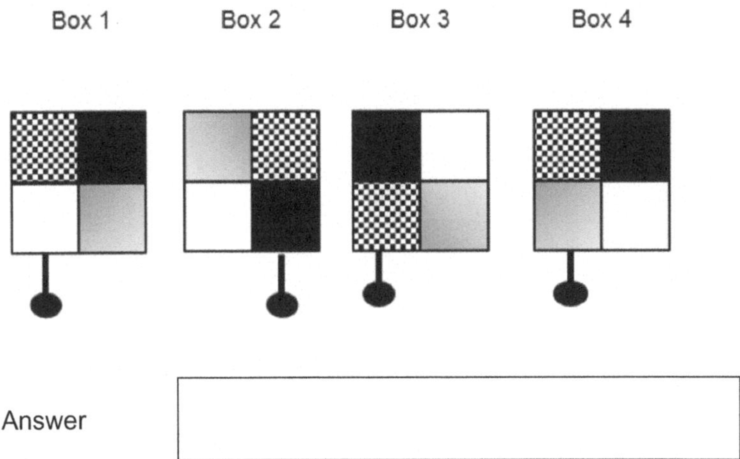

Answer

Question 10

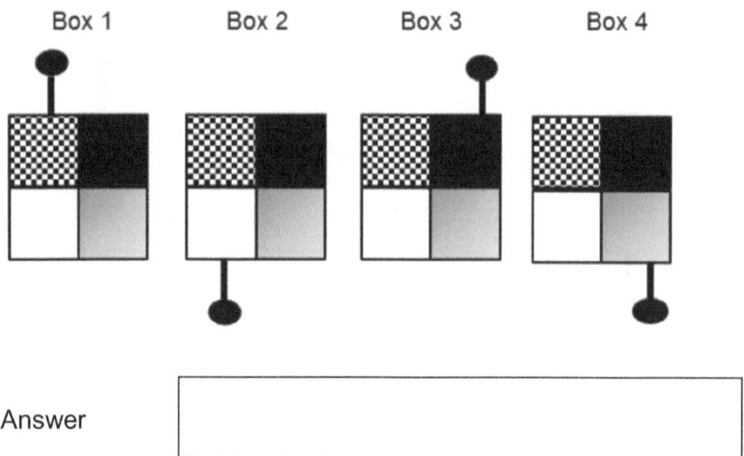

Answer

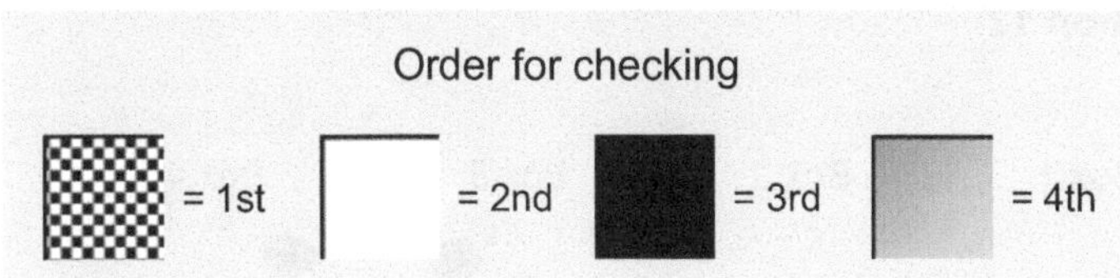

For the following ten questions, write down the order in which the boxes should be checked using the 'Order for Checking' sequence above:

Question 11

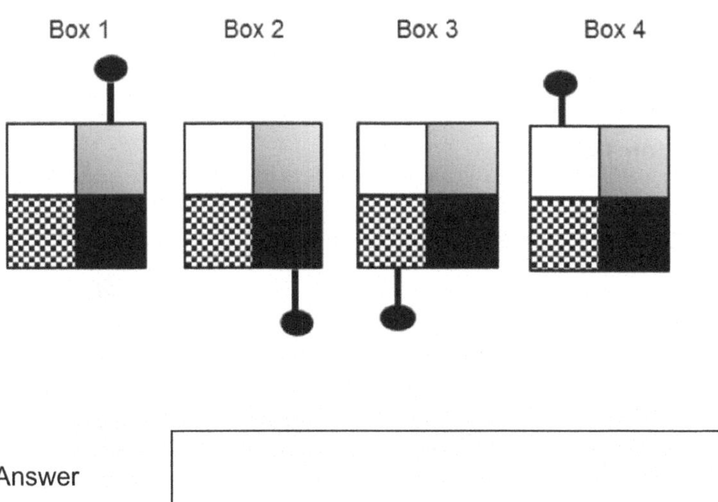

Answer

Question 12

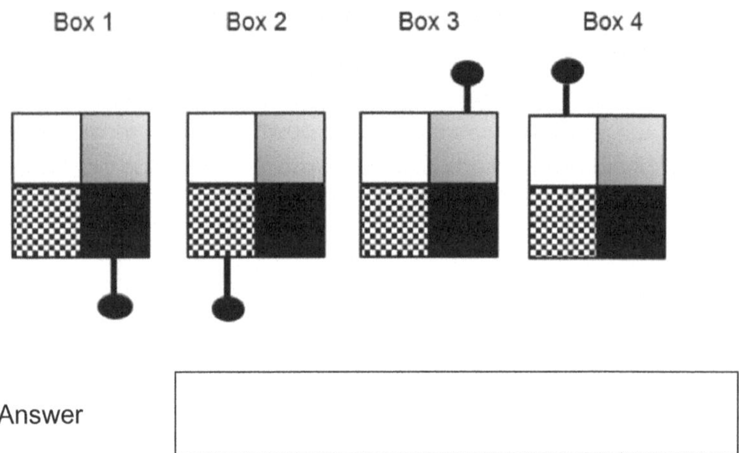

Answer

Question 13

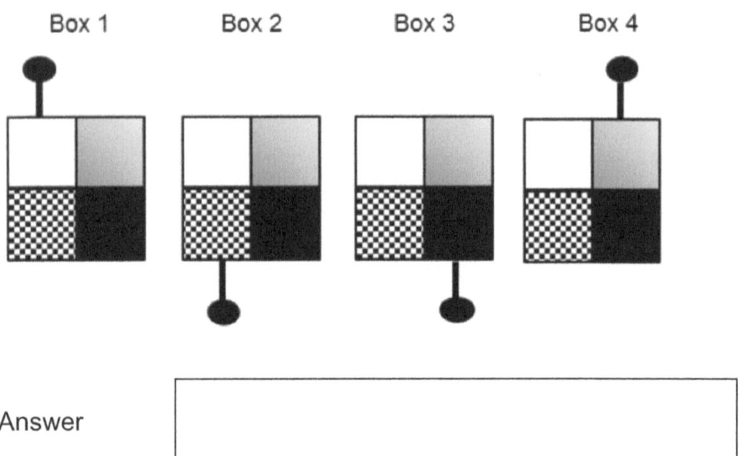

Answer

Question 14

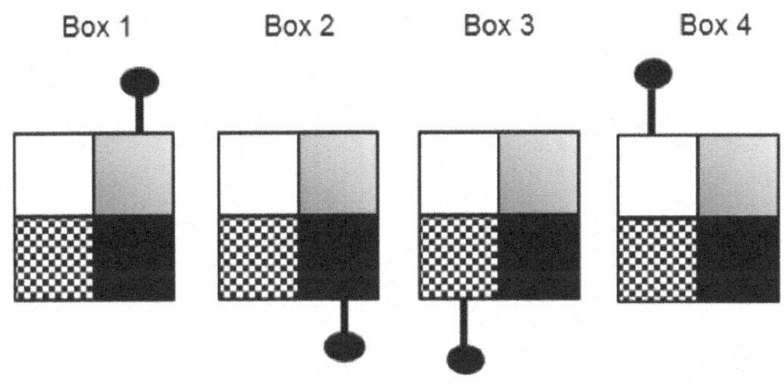

Answer

Question 15

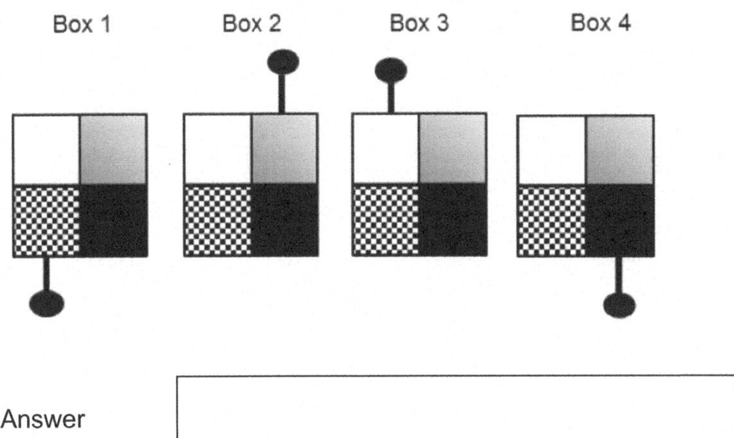

Answer

Question 16

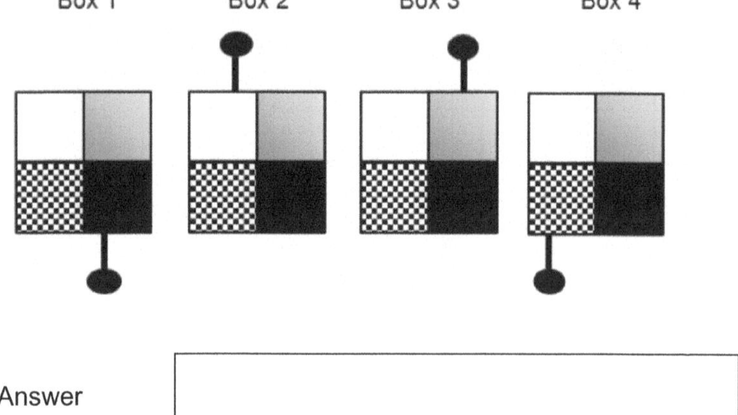

Answer

Question 17

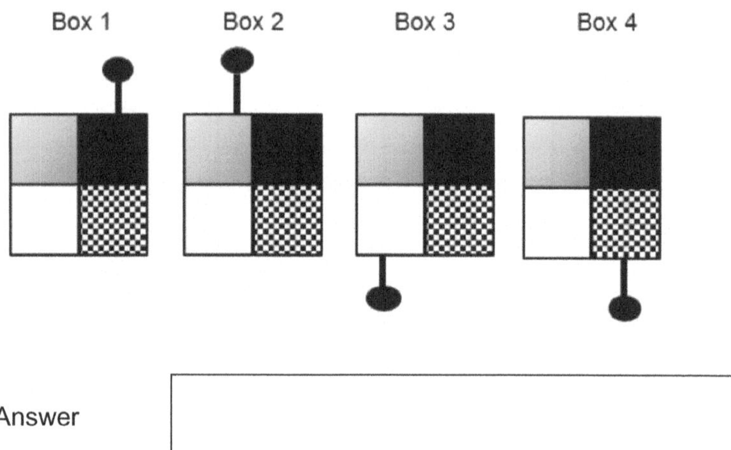

Answer

Question 18

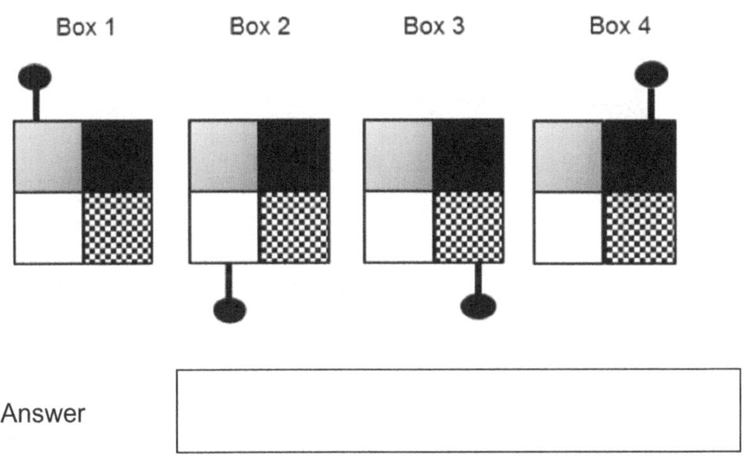

Answer

Question 19

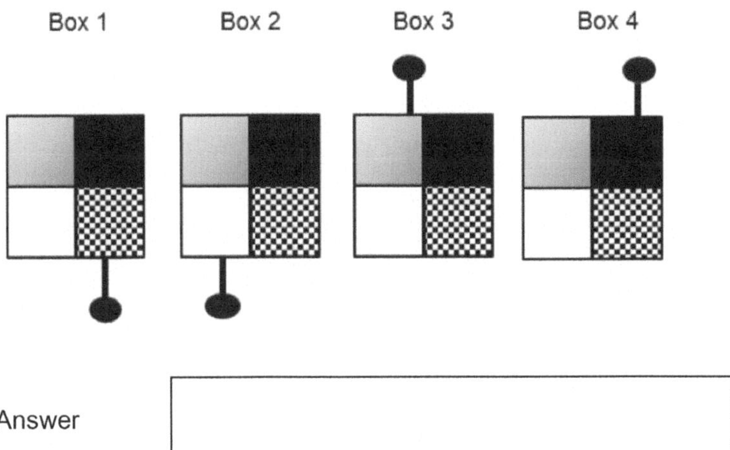

Answer

Question 20

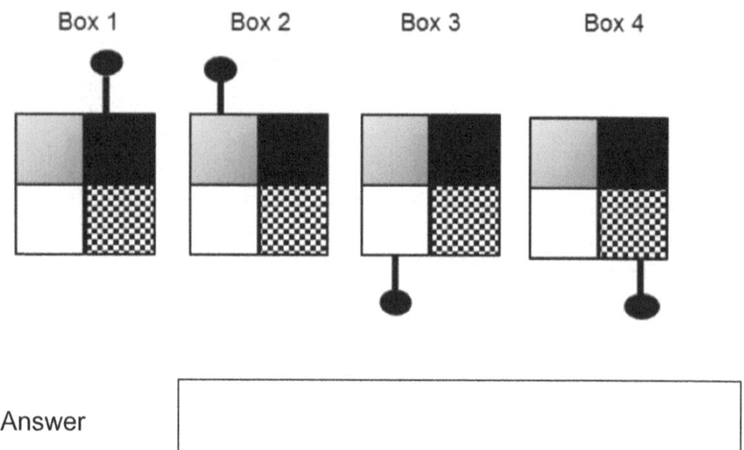

Answer

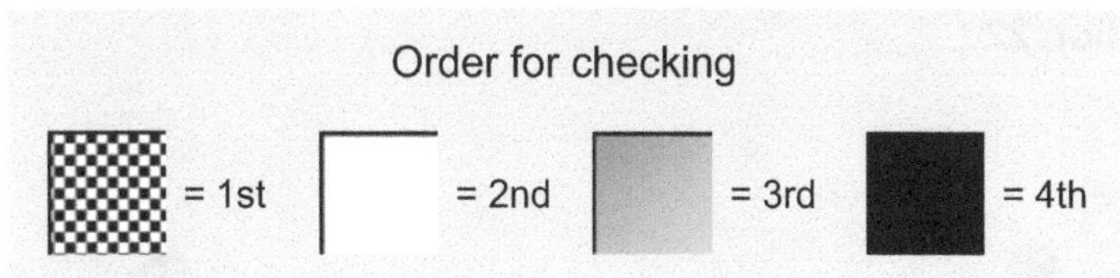

For the following five questions, write down the order in which the boxes should be checked using the 'Order for Checking' sequence above

Question 21

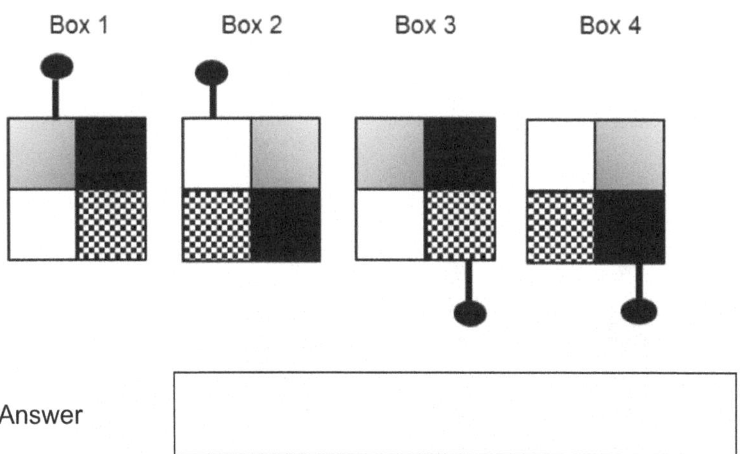

Answer

Question 22

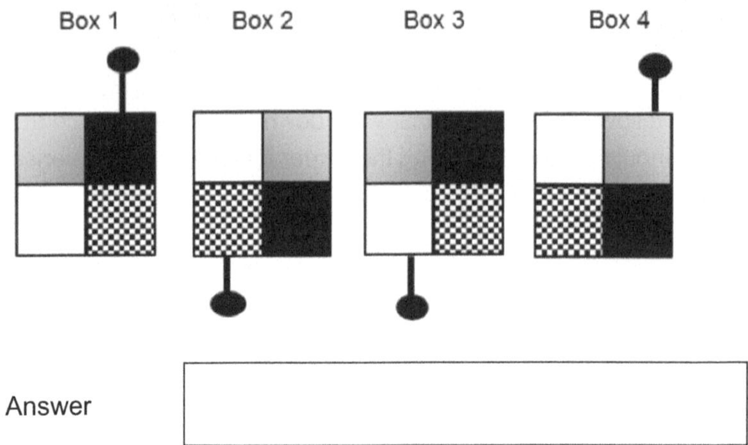

Answer

Question 23

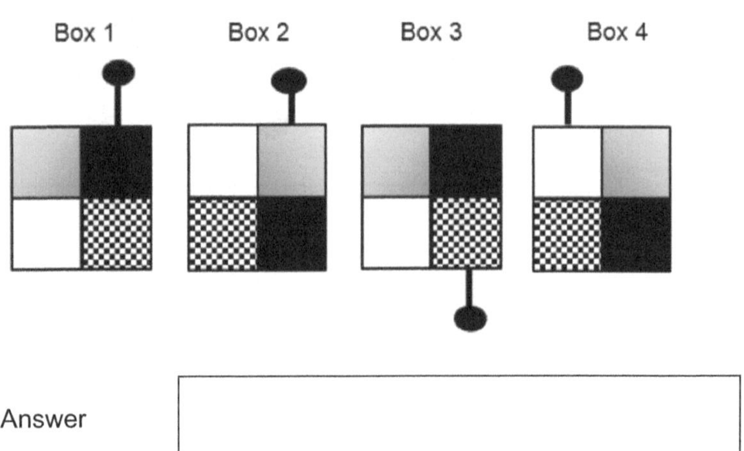

Answer

Question 24

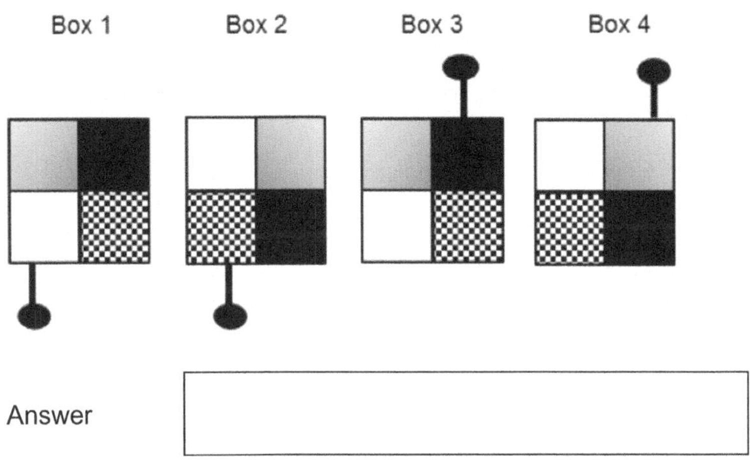

Answer

Question 25

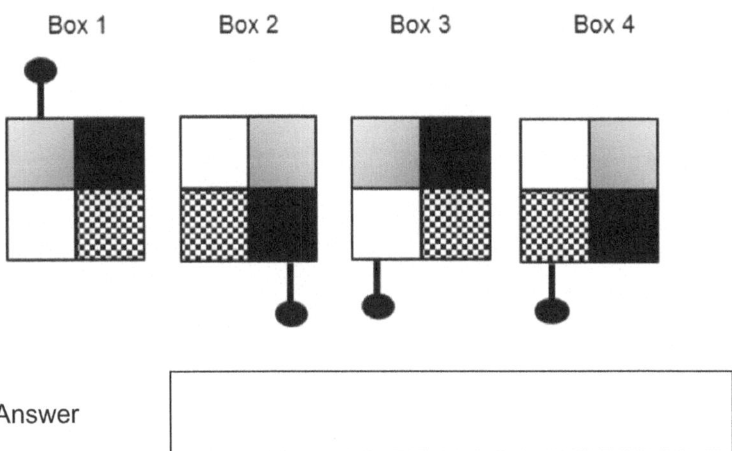

Answer

ANSWERS TO CHECKING TESTS – TEST SECTION 1

1.   3214
2.   3142
3.   1324
4.   3412
5.   4123
6.   2431
7.   1324
8.   1243
9.   1324
10. 2134
11. 3421
12. 2413
13. 2134
14. 3421
15. 1342
16. 4213
17. 4312
18. 3241
19. 1243
20. 4312
21. 3214
22. 2341
23. 3421
24. 2143
25. 4312

CHECKING TESTS – TEST SECTION 2

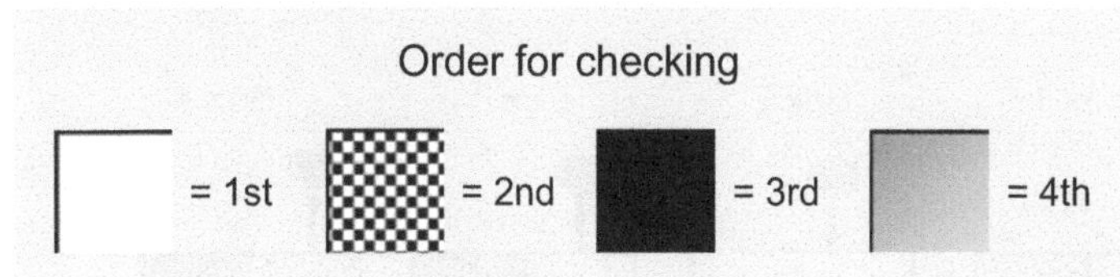

For the following ten questions, write down the order in which the boxes should be checked using the 'Order for Checking' sequence above:

Question 1

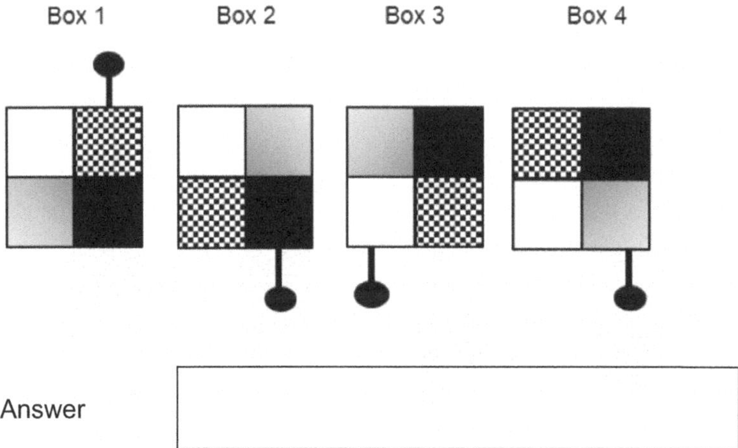

Answer

Question 2

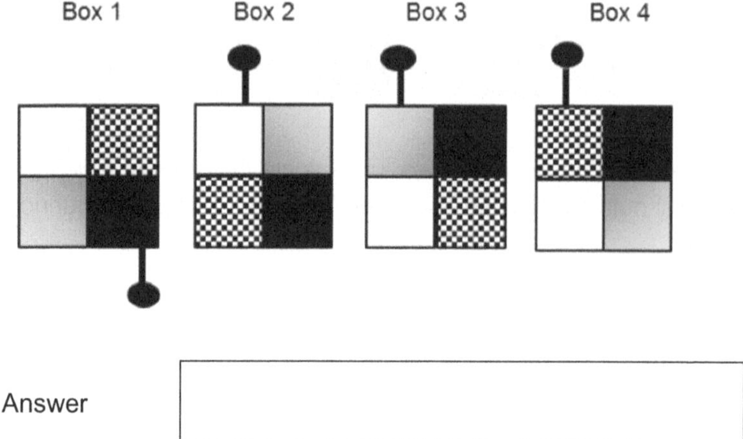

Answer

Question 3

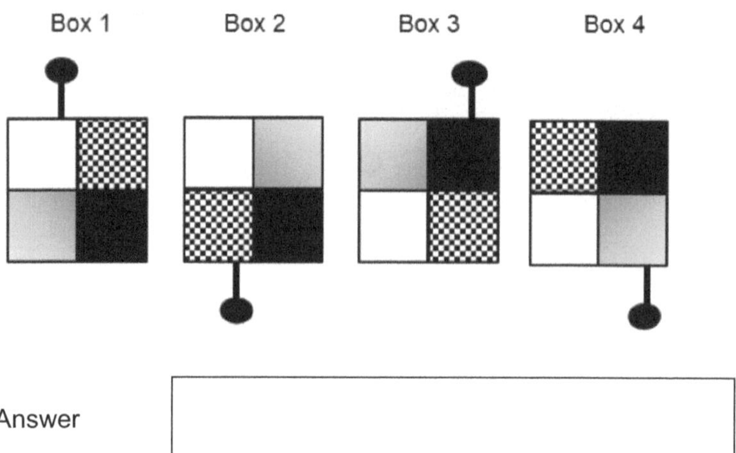

Answer

Question 4

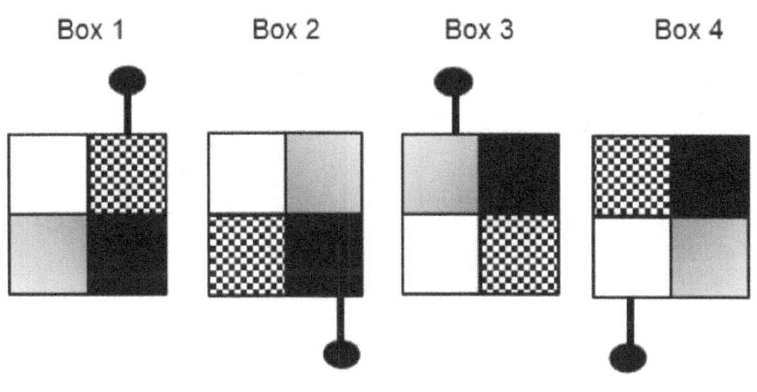

Answer

Question 5

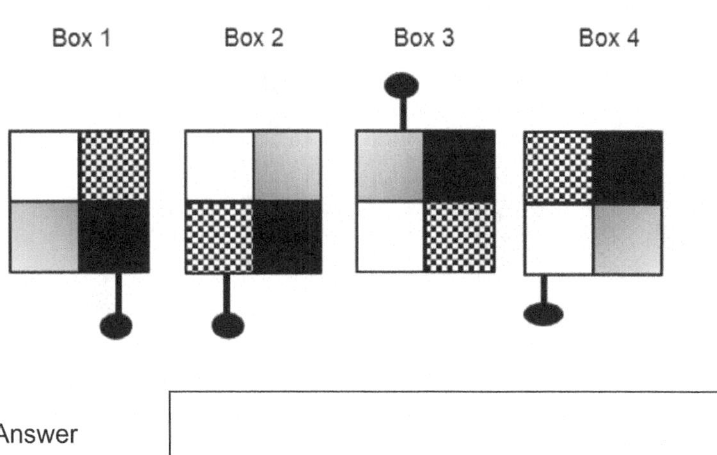

Answer

Question 6

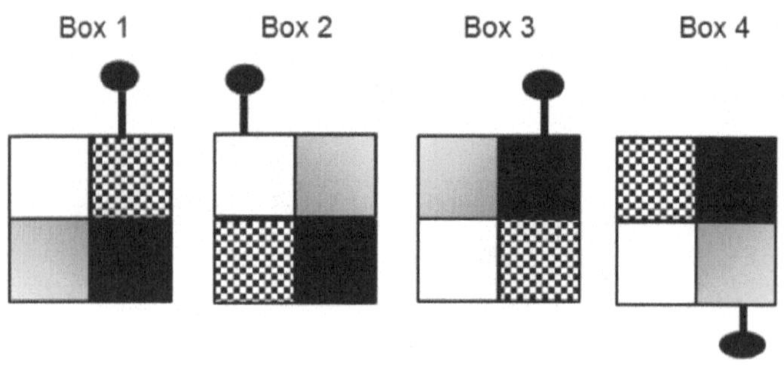

Answer

Question 7

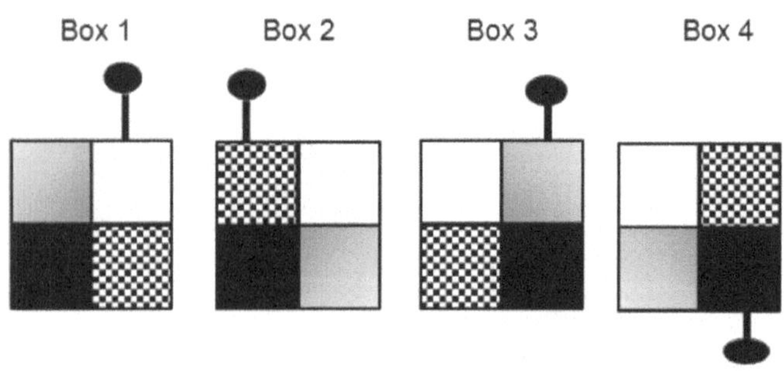

Answer

Question 8

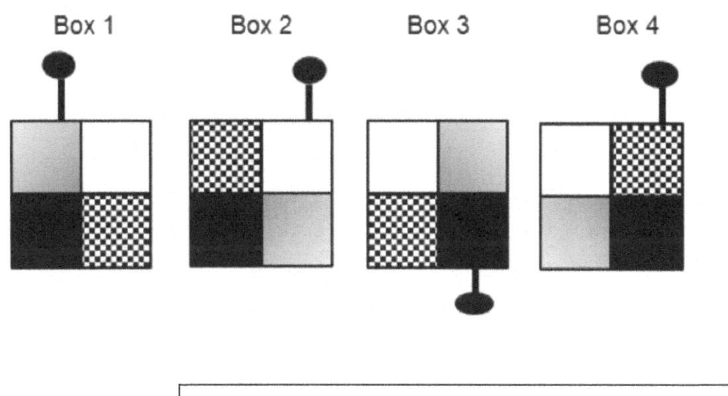

Answer

Question 9

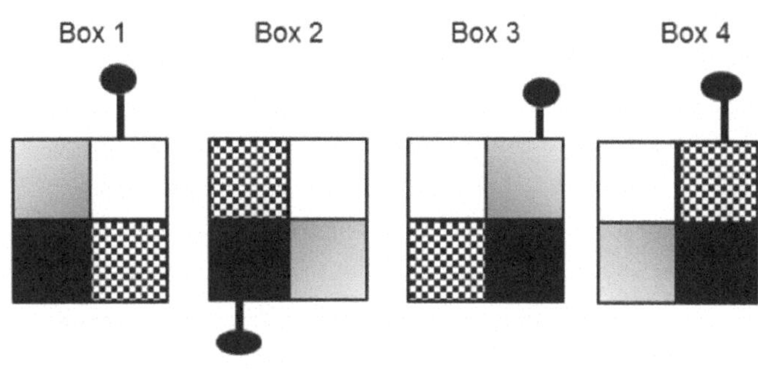

Answer

Question 10

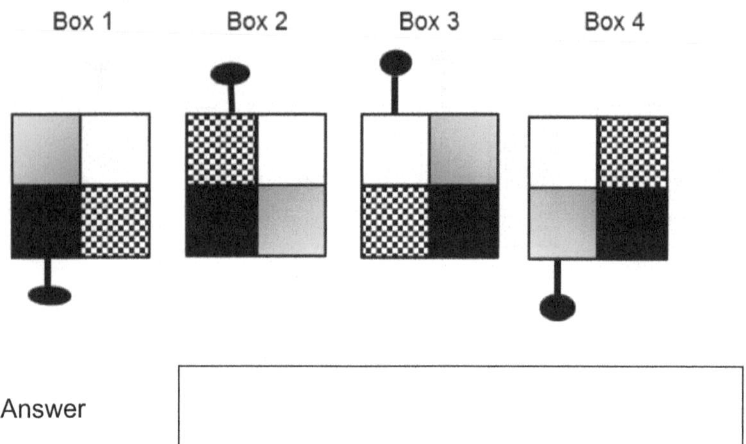

Answer

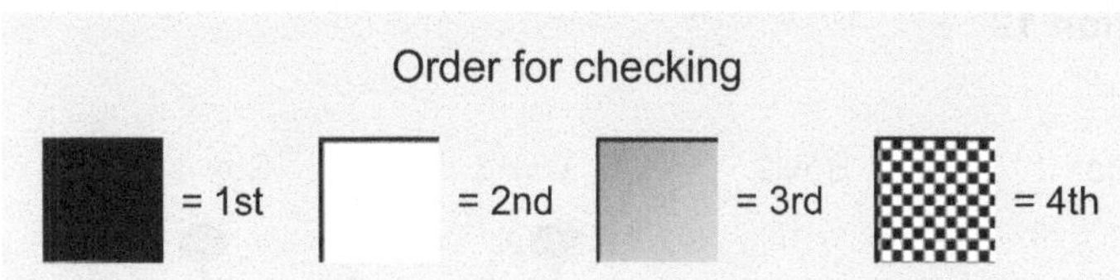

For the following ten questions, write down the order in which the boxes should be checked using the 'Order for Checking' sequence above:

Question 11

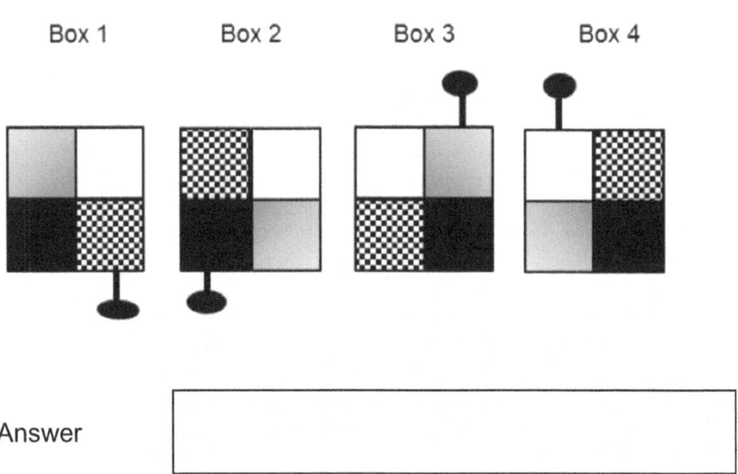

Answer

Question 12

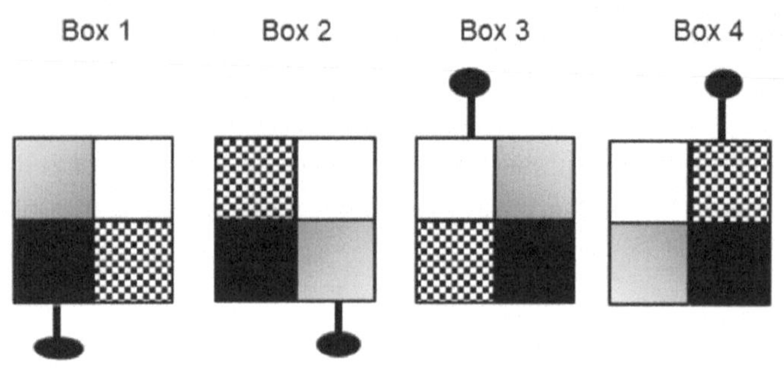

Answer

Question 13

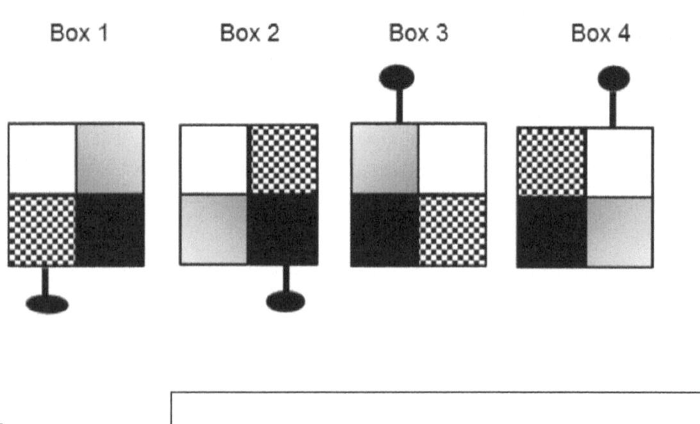

Answer

Question 14

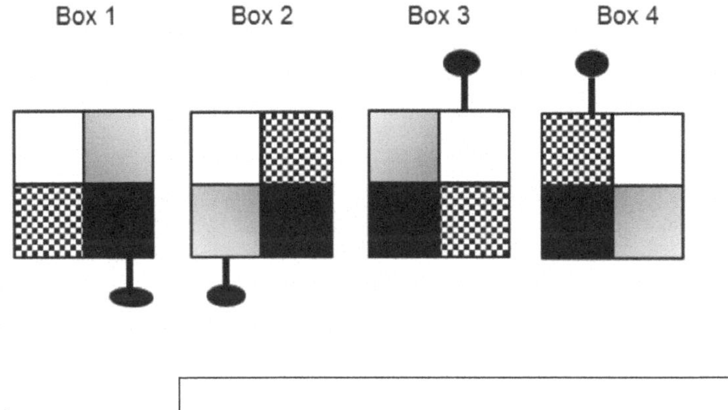

Answer

Question 15

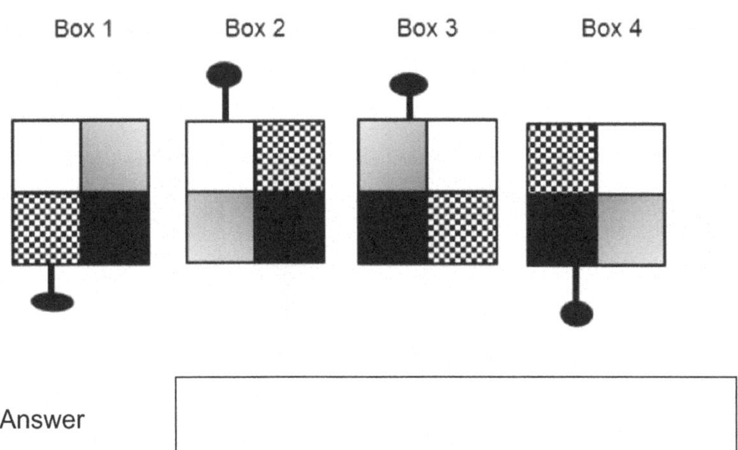

Answer

Question 16

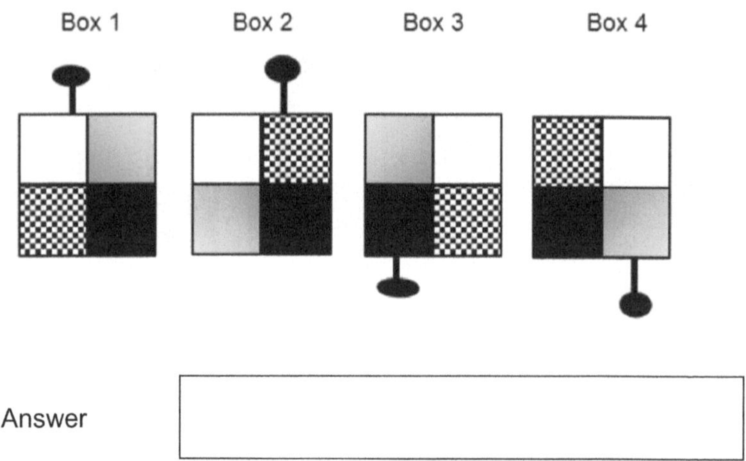

Answer

Question 17

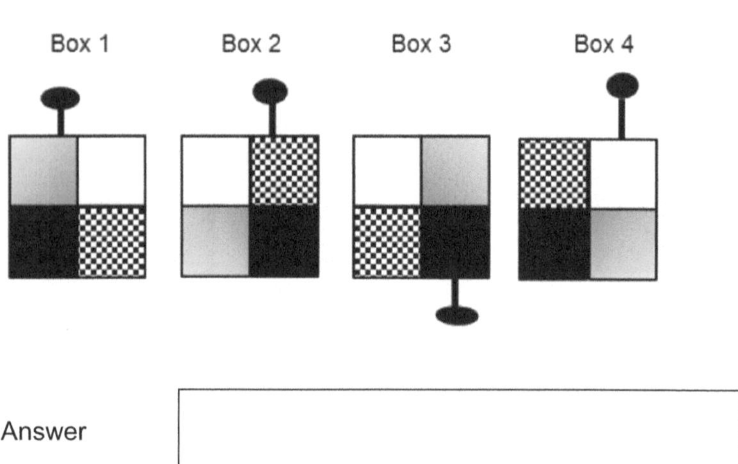

Answer

Question 18

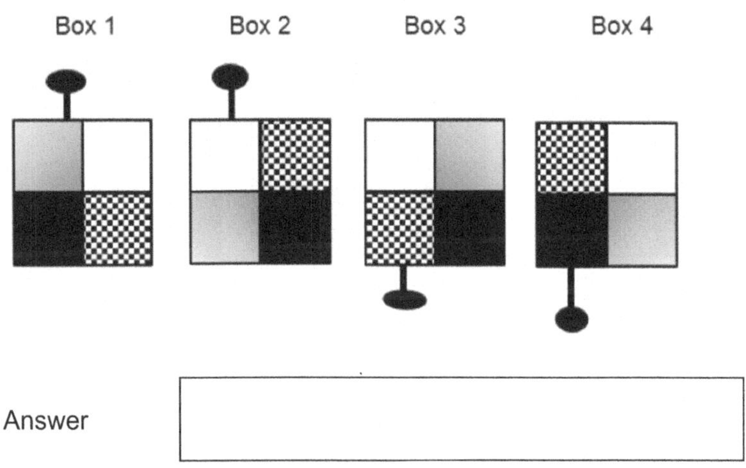

Answer

Question 19

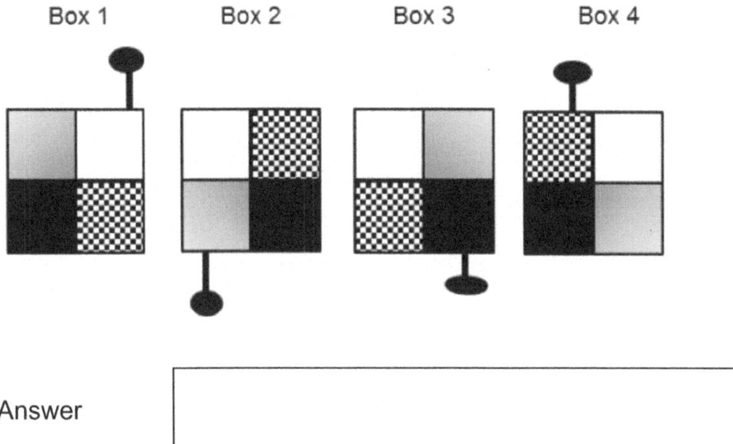

Answer

Question 20

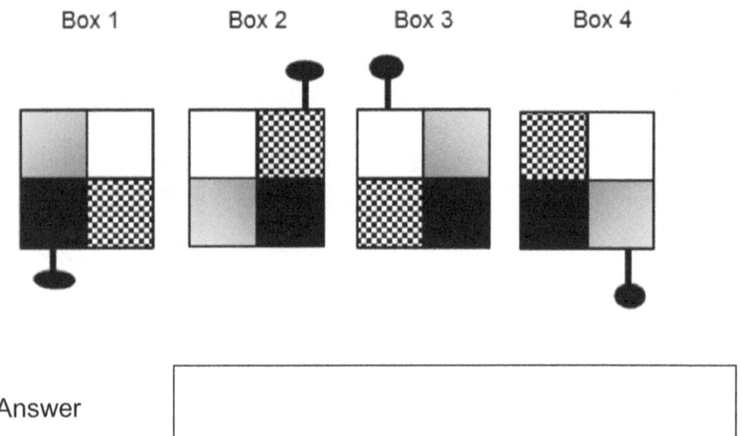

Answer

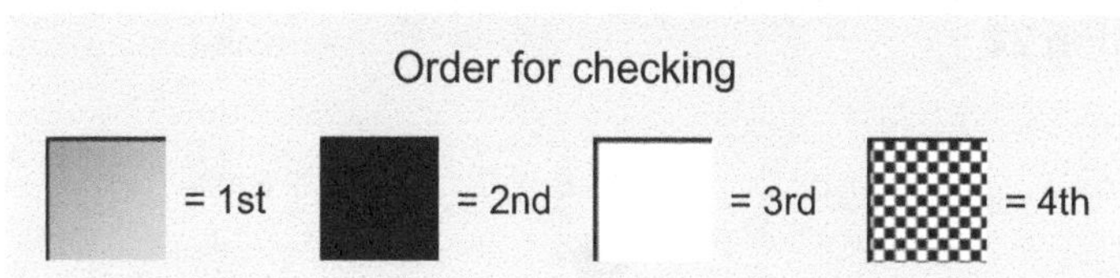

For the following five questions, write down the order in which the boxes should be checked using the 'Order for Checking' sequence above

Question 21

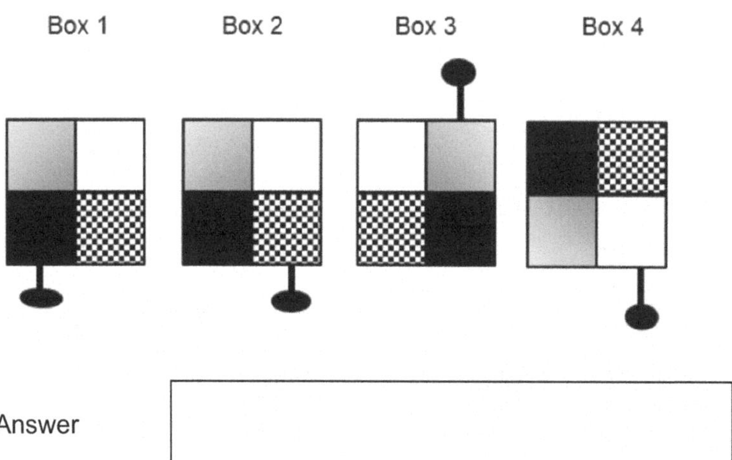

Answer

Question 22

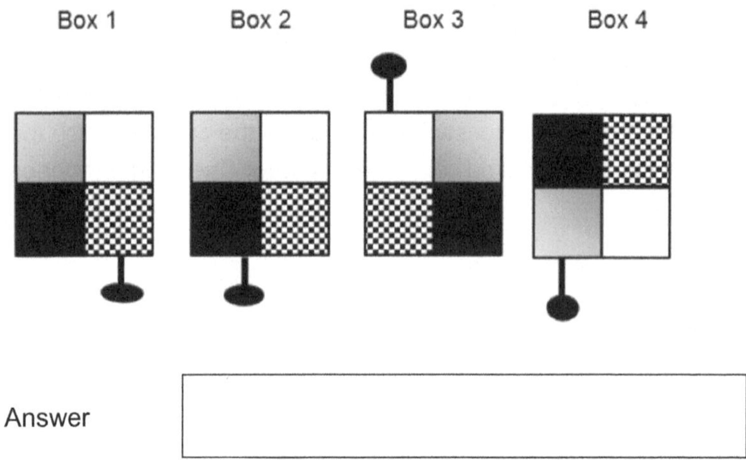

Answer

Question 23

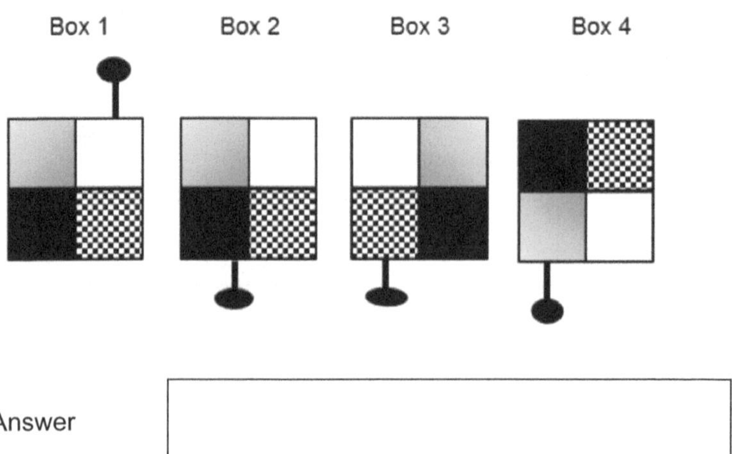

Answer

Question 24

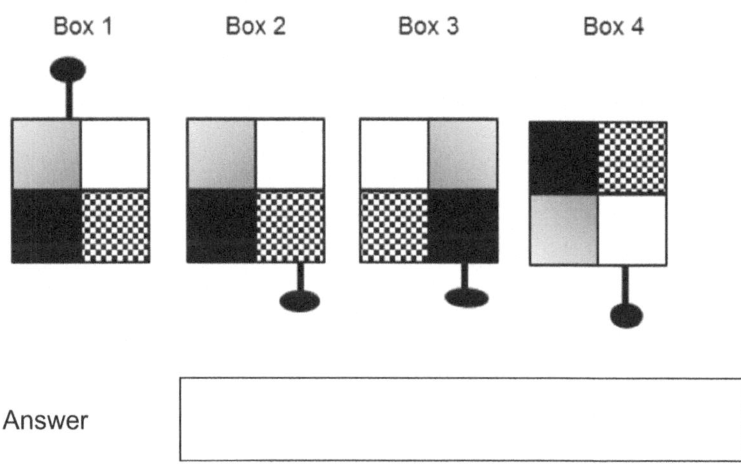

Answer

Question 25

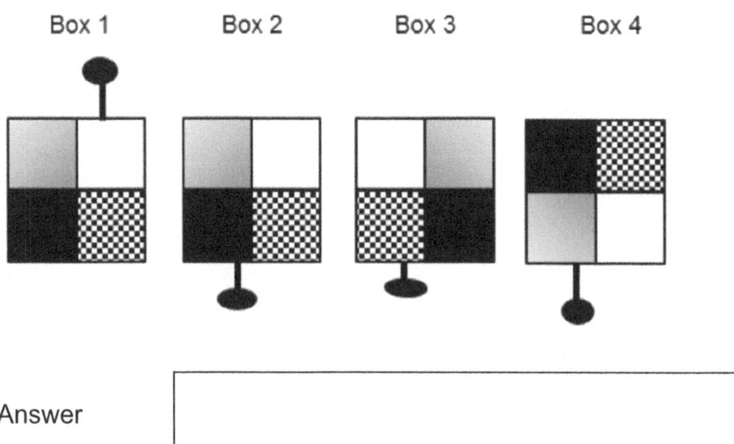

Answer

ANSWERS TO CHECKING TESTS – TEST SECTION 2

1. 3124
2. 2413
3. 1234
4. 4123
5. 4213
6. 2134
7. 1243
8. 2431
9. 1423
10. 3214
11. 2431
12. 1324
13. 2431
14. 1324
15. 4231
16. 3142
17. 3412
18. 4213
19. 3124
20. 1342
21. 3142
22. 4231
23. 4213
24. 1342
25. 4213

DOTS CONCENTRATION TESTS (DCT)

The Dots Concentration Test (DCT) is a test that assesses your ability to concentrate for long periods of time, and is probably the hardest skill to master in psychometric testing. Many candidates turn up to take the test without any prior knowledge of how it works and what is expected of them.

The test is designed to assess your ability to concentrate whilst performing tasks at high speed. You will be presented with five pages or screens that each contains 25 columns. Each of the columns contains boxes with patterns of dots. Your task is to work quickly and accurately through each column, from left to right, identifying boxes of 4 dots only.

You are allowed two minutes only per sheet and, once the two minutes are up, you are told to move onto the next page regardless of whether you have completed it or not. The test requires ten minutes of solid concentration.

Take a look at the following row of dots:

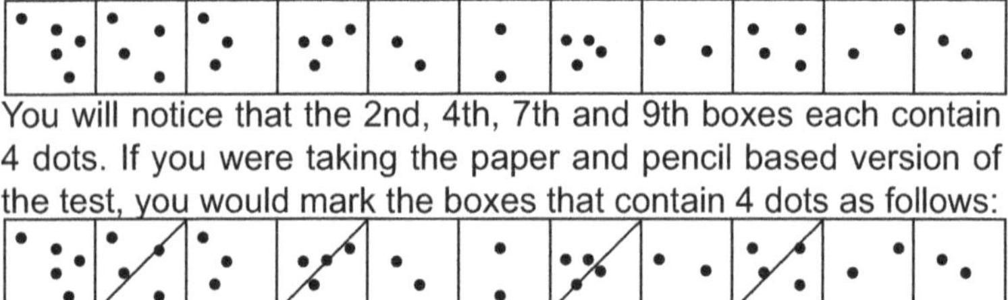

You will notice that the 2nd, 4th, 7th and 9th boxes each contain 4 dots. If you were taking the paper and pencil based version of the test, you would mark the boxes that contain 4 dots as follows:

You will notice that I have placed a single diagonal line through each of the boxes that contains 4 dots.

If you are required to undertake the computer based version of the test then you will be required to use the keys on the keyboard as follows:

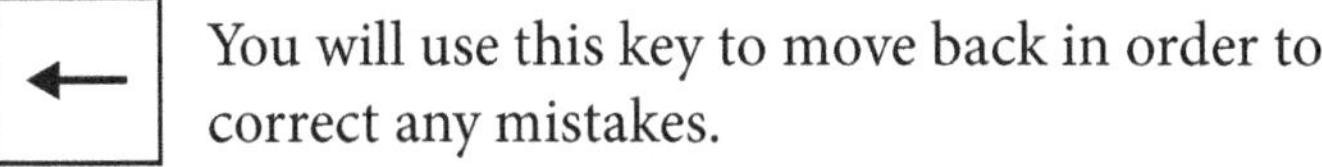

You will use this key to move from left to right across the screen.

You will use this key to mark each box that contains 4 dots.

You will use this key to move back in order to correct any mistakes.

On the following pages I have provided you with ten sample concentration tests. During the first set of five concentration tests you are required to locate specific letters and/or numbers that are contained within rows and columns. Full instructions are provided at the start of each test.

During the second set of five tests you will be required to search for groups of 4 dots in rows and columns of boxes. Once again, full instructions are provided.

SAMPLE CONCENTRATION TEST 1

Cross out the letter 'R' (upper case) in each row. Write down the total number that you cross out in each row in the box provided at the end of each row. You have 60 seconds to complete the test.

1.	Q	r	R	g	y	U	h	J	R	j	R	k	L	B	n	
2.	R	R	R	v	B	n	M	U	u	d	f	O	p	T	R	
3.	C	x	X	F	R	G	t	p	A	R	f	V	R	y	U	
4.	Q	R	R	t	G	N	H	J	r	r	F	P	F	R	r	
5.	Q	a	Z	x	R	t	I	o	M	B	R	D	x	A	S	
6.	R	s	a	A	e	E	R	C	Y	U	r	j	P	o	R	
7.	T	R	r	P	F	r	S	N	b	V	c	F	F	R	R	
8.	G	v	R	r	R	y	R	P	R	r	D	e	E	R	F	
9.	T	R	K	P	o	u	b	g	t	m	R	r	X	r	R	
10.	C	B	n	h	j	Y	I	p	R	R	R	r	R	C	d	
11.	R	R	r	Y	u	B	v	M	n	h	K	j	R	E	R	
12.	A	W	r	E	R	f	p	U	I	H	R	y	U	B	R	
13.	R	r	Q	q	B	G	R	t	Q	w	E	F	T	y	R	
14.	T	R	A	I	N	D	P	I	V	E	R	D	T	y	S	
15.	d	x	z	Z	R	n	K	i	i	R	r	R	O	p	o	
16.	Q	R	r	E	D	D	e	w	K	i	I	O	P	R	R	
17	H	O	w	B	e	E	R	r	R	R	V	R	H	j	R	
18.	K	j	u	U	Y	i	Y	r	R	R	D	X	z	q	Q	
19.	P	y	g	h	j	I	r	t	r	e	R	e	R	q	Z	
20.	B	h	B	h	r	r	R	r	N	B	H	y	Y	R	F	

SAMPLE CONCENTRATION TEST 2

Cross out the letter 'o' (lower case). Write down the total number that you cross out in each row in the box provided at the end of each row. You have 60 seconds to complete the test.

1.	o	O	t	Q	w	q	O	o	A	B	u	U	o	o	O	
2.	O	o	g	Y	t	B	c	C	c	O	o	o	o	D	w	
3.	B	o	O	g	a	s	S	q	Q	t	Q	q	O	o	G	
4.	I	L	N	h	U	u	O	o	H	y	t	R	o	O	o	
5.	G	V	v	R	t	Y	o	o	P	i	O	O	o	O	R	
6.	G	t	y	U	J	P	p	O	o	D	d	O	o	S	Q	
7.	O	o	O	o	o	o	Y	t	Y	q	Q	q	o	c	c	
8.	I	u	V	c	c	F	r	d	w	H	y	h	u	o	o	
9.	Y	o	o	U	o	O	O	y	D	e	q	A	q	O	o	
10.	R	r	t	o	u	y	G	b	t	r	e	o	o	o	P	
11.	o	O	c	o	d	d	D	O	c	c	O	o	o	d	R	
12.	B	v	c	f	R	o	y	f	D	r	d	r	a	A	a	
13.	F	t	t	t	d	r	e	o	o	p	u	o	Q	t	r	
14.	F	g	r	t	y	N	H	N	h	o	p	O	o	I	y	
15.	T	r	e	d	w	o	u	i	y	F	c	r	D	e	W	
16.	o	o	O	o	p	O	u	i	S	t	d	r	s	S	O	
17	I	o	O	A	a	a	c	C	c	g	o	o	o	R	t	
18.	G	g	g	g	o	t	f	d	r	t	u	u	o	o	j	
19.	Q	c	v	b	g	t	y	u	O	o	O	o	G	y	c	
20.	K	I	o	i	u	y	t	r	e	o	u	y	o	j	h	

SAMPLE CONCENTRATION TEST 3

Cross out the letters 'w' (lower case) and 'V' (upper case). Search for both of these letters at the same time. Write down the total combined number that you cross out in each row in the box provided at the end of each row. You have 60 seconds to complete the test.

1.	v	W	w	V	e	w	h	j	U	i	X	x	W	w	v
2.	V	u	U	w	G	t	y	u	W	w	V	v	W	o	o
3.	W	W	V	V	v	v	w	w	y	u	i	p	v	W	W
4.	V	g	h	j	K	O	p	t	Y	V	v	W	W	w	V
5.	Y	U	u	u	v	v	W	M	m	w	e	V	v	N	n
6.	q	q	Q	G	g	H	Y	u	i	R	T	y	V	w	v
7.	V	y	u	Y	u	o	p	N	h	j	W	w	V	V	v
8.	t	y	m	k	m	N	b	C	x	W	w	V	v	b	v
9.	O	o	V	v	f	g	h	j	k	n	h	N	h	V	X
10.	T	V	v	X	c	d	W	w	W	v	V	v	f	r	p
11.	V	V	v	w	W	w	v	V	v	W	w	g	y	Y	v
12.	R	t	y	u	i	B	g	v	f	r	D	r	Q	w	W
13.	R	t	y	V	c	V	c	v	f	r	W	w	W	w	V
14.	G	y	u	i	O	p	R	t	y	E	w	V	V	v	W
15.	Y	Y	y	Y	X	v	W	W	w	w	r	t	y	u	v
16.	W	w	w	v	t	u	i	n	h	v	V	w	W	w	f
17	r	t	y	y	u	i	V	b	n	h	g	w	w	W	w
18.	i	o	q	w	S	S	X	W	V	Z	z	V	v	W	y
19.	P	o	Y	u	i	V	v	X	w	W	w	R	t	R	y
20.	y	u	V	x	s	t	Y	u	y	W	w	C	d	V	w

SAMPLE CONCENTRATION TEST 4

Cross out the number 8 and the letter 'b' (lower case). Search for both letter and number at the same time. Write down the total combined number that you cross out in each row in the box provided at the end of each row. You have 60 seconds to complete the test.

1.	8	B	8	V	v	W	q	P	p	r	g	B	b	8	u	
2.	B	b	R	r	r	y	U	i	8	8	B	B	b	g	G	
3.	j	u	p	P	b	v	f	r	B	b	w	3	6	7	R	
4.	8	3	2	h	y	U	x	W	w	v	x	v	b	B	8	
5.	f	G	g	B	p	h	b	b	b	B	B	8	8	5	3	
6.	y	u	U	7	6	5	8	e	r	d	r	w	8	B	b	
7.	o	O	o	P	7	8	5	b	3	8	3	R	r	S	l	
8.	B	b	3	8	B	B	b	h	h	V	c	b	B	7	1	
9.	1	3	c	V	f	l	u	y	t	r	B	b	8	8	8	
10.	y	B	b	8	4	3	3	3	X	x	x	f	F	r	t	
11.	Q	q	H	b	B	b	8	B	6	3	3	2	u	B	b	
12.	G	G	g	B	b	8	3	8	3	D	d	D	l	P	p	
13.	G	b	b	8	8	6	5	4	0	L	o	P	p	P	B	
14.	3	B	b	8	3	B	B	b	3	E	e	3	8	4	P	
15.	t	Y	y	D	e	e	D	f	g	W	8	8	P	P	B	
16.	C	C	b	n	B	8	B	8	B	b	8	3	9	3	9	
17	6	6	b	B	8	8	d	k	l	p	o	U	S	y	Y	
18.	P	p	8	F	d	D	c	C	8	B	b	8	f	F	f	
19.	8	8	C	f	z	s	W	w	R	r	T	8	3	B	b	
20.	H	y	y	b	B	8	8	8	H	H	h	D	r	e	W	

SAMPLE CONCENTRATION TEST 5

Cross out the letter 'e' (lower case) and the number '3'. Search for both letter and number at the same time. Write down the number crossed out in the box provided at the end of each row. You have 60 seconds to complete the test.

1.	E	6	e	8	8	e	3	p	b	d	e	E	3	8	T	
2.	e	8	3	6	7	y	u	I	V	f	E	e	b	B	E	
3.	W	w	q	D	d	c	x	z	O	p	e	R	6	8	3	
4.	y	u	I	o	p	P	t	T	Y	e	E	3	8	6	F	
5.	g	B	4	3	2	7	8	3	e	E	3	4	E	e	3	
6.	e	3	3	e	E	d	W	q	h	j	K	8	7	N	9	
7.	3	e	E	8	B	8	3	e	E	k	K	3	e	8	7	
8.	f	C	x	b	g	t	T	r	6	8	3	4	X	d	e	
9.	3	3	3	b	8	b	e	3	E	3	8	3	4	0	1	
10.	e	E	j	H	g	b	3	E	e	3	w	b	V	v	E	
11.	8	3	B	v	C	f	v	e	8	4	3	3	3	e	v	
12.	6	7	8	v	c	D	f	3	7	8	6	E	e	e	V	
13.	e	3	e	3	E	8	E	3	e	E	3	2	8	G	g	
14.	7	y	h	n	g	f	d	e	E	4	E	e	3	D	d	
15.	k	I	L	j	h	y	V	v	8	4	2	b	V	v	E	
16.	g	Y	y	i	9	8	7	0	3	O	o	v	V	v	e	
17	8	2	B	b	v	e	W	e	r	5	5	R	r	e	V	
18.	3	e	E	e	3	4	b	V	v	e	W	w	q	A	a	
19.	5	e	3	V	f	r	6	5	4	e	e	E	e	3	E	
20.	e	E	e	R	3	4	2	1	3	E	e	h	G	f	d	

ANSWERS TO CONCENTRATION TESTS

Row #	Test 1	Test 2	Test 3	Test 4	Test 5
1.	3	4	4	4	5
2.	4	4	4	4	3
3.	3	2	4	2	2
4.	3	3	4	3	2
5.	2	3	2	5	6
6.	3	2	2	3	4
7.	3	5	4	3	6
8.	5	2	2	4	2
9.	3	4	2	4	7
10.	4	4	3	2	4
11.	4	4	6	4	6
12.	3	1	1	3	3
13.	3	3	5	4	7
14.	2	2	3	4	3
15	3	1	2	2	0
16.	3	3	5	5	2
17.	5	4	4	3	3
18.	2	3	3	4	5
19.	2	2	3	4	6
20.	2	3	4	4	5

Check through your answers carefully and go back to check over the ones you got wrong.

Now move onto to the next set of five tests.

SAMPLE TEST 1

Place a diagonal line across each box that contains 4 dots only. You have 30 seconds to complete the test.

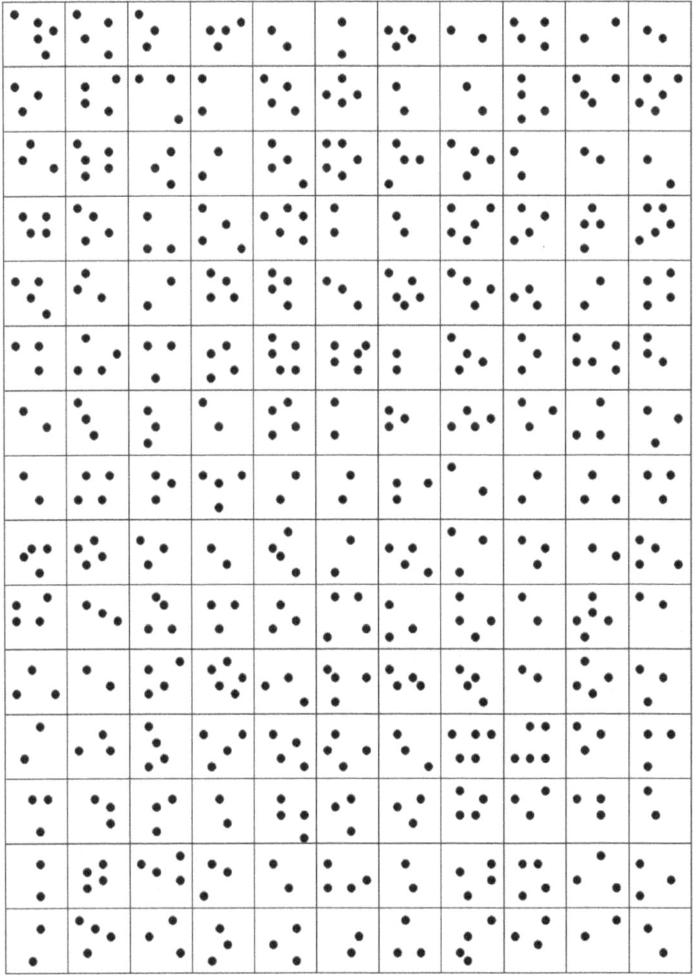

SAMPLE TEST 2

Place a diagonal line across each box that contains 4 dots only.
You have 30 seconds to complete the test.

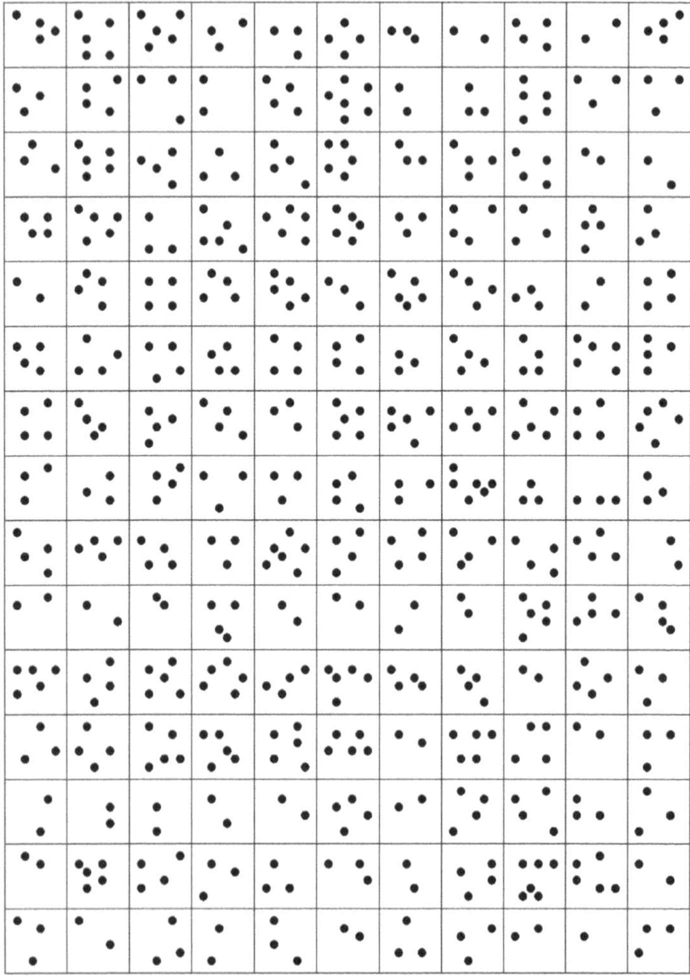

SAMPLE TEST 3

Place a diagonal line across each box that contains 4 dots only.
You have 30 seconds to complete the test.

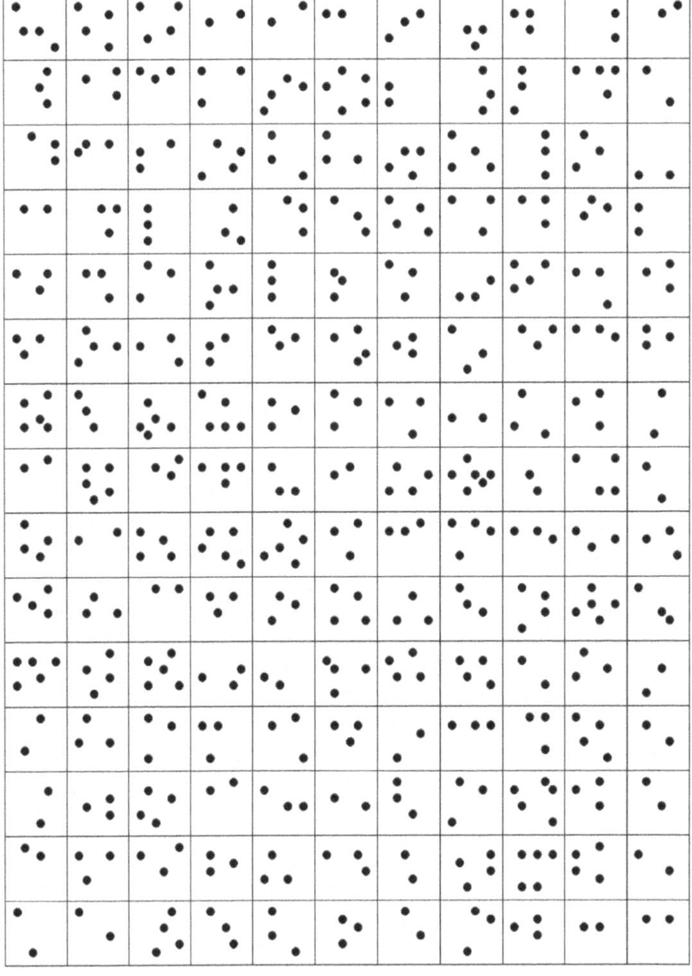

SAMPLE TEST 4

Place a diagonal line across each box that contains 4 dots only. You have 30 seconds to complete the test.

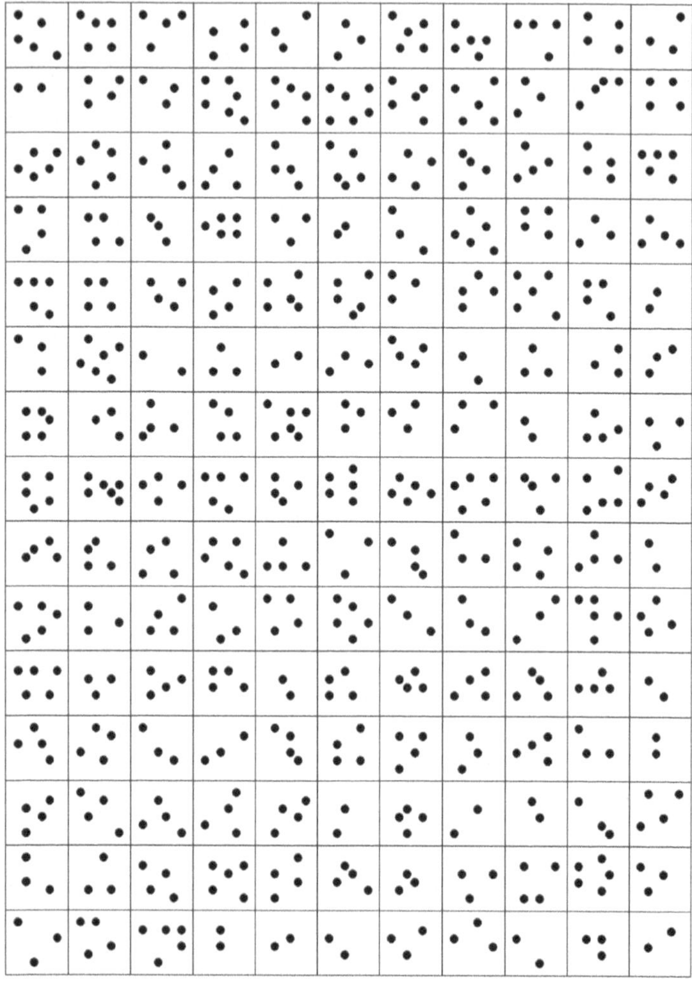

SAMPLE TEST 5

Place a diagonal line across each box that contains 4 dots only. You have 30 seconds to complete the test.

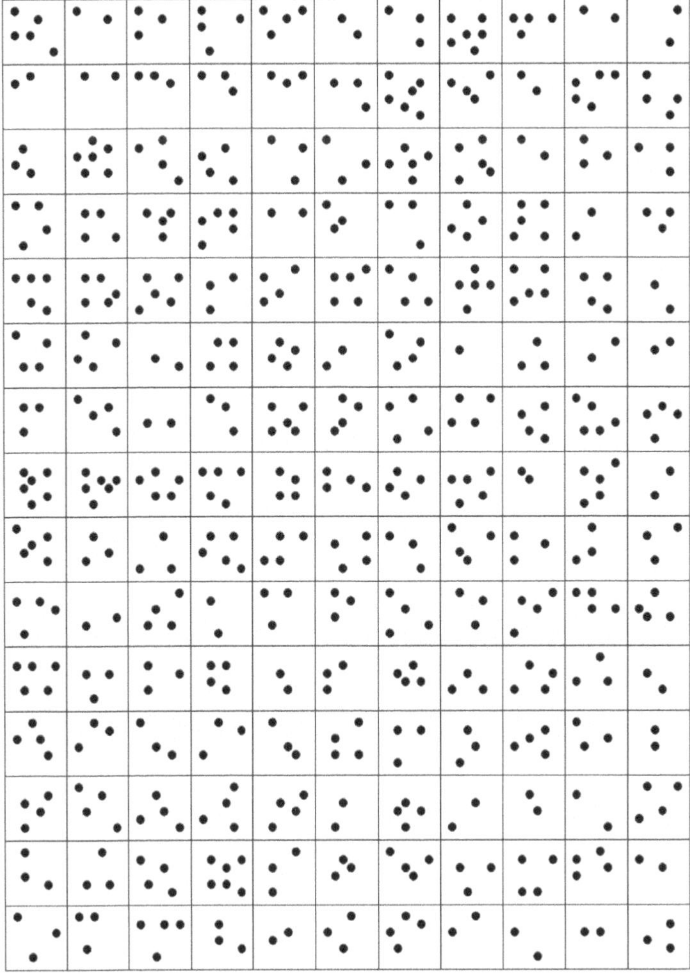

ANSWERS TO DOTS CONCENTRATION TESTS 1 TO 5

Test 1 - 56 boxes containing groups of 4 dots

Test 2 - 58 boxes containing groups of 4 dots

Test 3 - 31 boxes containing groups of 4 dots

Test 4 - 66 boxes containing groups of 4 dots

Test 5 - 56 boxes containing groups of 4 dots

THE INTERVIEW

During this section of the guide I will provide you with some useful tips on how to prepare for the Train Driver interviews. The information provided will be useful for both the structured interview and also the manager interview.

The Train Driver Interview does not have to be a daunting process, providing that you prepare effectively. Yes, any interview can be a nerve-wracking experience, but if you prepare in the right areas this will give the confidence you need to pass with flying colours. Within this section of the guide I have provided you with a number of sample questions that you may get asked during your Interview. The structured interview may be carried out on the same day as the psychometric testing, so make sure you have prepared for it well in advance. The panel will normally consist of at least one Train Driver Manager and a representative from the Human Resources department. The Human Resources representative is present to ensure that the interview is carried out fairly and in conjunction with company policy and guidelines. Every candidate will be asked the same questions to ensure consistency and fairness.

HOW TO PREPARE EFFECTIVELY

During your preparation for the interview I would recommend that you concentrate on the following three key areas:

- Interview technique;
- Research;
- Responding to the interview questions.

Each of the above areas is equally important. I will now go into each one of them in detail:

INTERVIEW TECHNIQUE

Interview technique covers a number of different areas. The majority of candidates will pay it little, if any attention at all. Interview technique basically involves the following key areas:

- Creating the right impression. When you walk into the interview room you should stand up tall, smile and be polite and courteous to the panel. Do not sit down in the interview chair until invited to do so.

- Being presentable. During my time as an interviewer for a number of different jobs I have been amazed at the number of people who turn up inappropriately dressed. I have seen people turn up for interviews in jeans, t shirts and trainers! I strongly advise that you take the time to look smart and presentable. Remember you are applying to join an organisation that requires you to wear a uniform. If you dress smart and formal for the interview then you are far more likely to wear your uniform with pride.

- Sitting in the chair. The interview could last for up to an hour, depending on the length of your responses to the questions. This is a long time to concentrate for. Whilst in the interview chair sit up right at all times and never slouch.

- Motivation. Throughout the duration of the interview demonstrate a high level of motivation and enthusiasm. You do not want to come across as desperate, but conversely you must come across as highly motivated and determined to be successful. Always smile and be respectful of the interview panel.

- Communication. When communicating with the interview panel look them in the eye. This shows a level of confidence. You should also communicate in a clear and concise manner where possible.

- Asking questions. At the end of the interview you will be given the opportunity to ask questions. This is where some candidates let themselves down with silly or inappropriate questions that relate to leave or sick pay. It is quite acceptable to ask a couple of questions, however, keep them simple and relevant. Examples of good questions to ask are:

  Q. If I am successful, how long would it be before I start my training?

  Q. I have been looking into your company and I have been impressed with the 'meet the manager's' scheme that you operate for your customers. Has this been successful?

- A final parting statement. Once the interview has finished and you have asked your questions, you may wish to finish off with a final statement. Your final statement should say something about your desire and passion for becoming a Trainee Train Driver. The following is a good example of a final statement:

"I would like to say thank you for giving me the opportunity to be interviewed for the post today. Over the last few months I have been working hard to learn about the role and also about your company. If I am successful then I promise you that I will work hard to pass the tests and exams and I will be a loyal and professional employee of your team. Thank you."

RESEARCH

As you can imagine, in the build up to the interview you will need to carry out plenty of research. Research that is, in relation to the role of a Trainee Train Driver and also the Train Operating Company that you are applying to join. Here is a list of the more important areas I recommend that you study:

- The job description and person specification for the job that you are applying for.

- Your application form and the responses that you provided to all of the questions.

- The website of the Train operating Company you are applying to join. What is their customer service charter? Do they have a mission statement? What services do they provide? What is their geographical area? How many people work for them? Who is the person in charge? What stations do they operate out of? What trains do they operate? Do they operate any schemes in order to improve customer service? What are the future plans of the TOC?

- Try to visit a train station that the TOC operates out of. Speak to some of the staff at the station and ask them questions about the role they perform. Try to find out as much as possible about the TOC you are applying for. If you get the opportunity, speak to a qualified Train Driver who works for the TOC. You may also decide to telephone the TOCs Human Resources department and ask if you can go along to find out a little bit more about their organisation and what they expect from their employees.

RESPONDING TO THE INTERVIEW QUESTIONS

If I were preparing for the Trainee Train Driver interview right now, I would take each area of the role individually and prepare a detailed response setting out where I meet the requirements of it. Your response to each question that relates to the role of a Trainee Train Driver must be 'specific' in nature. This means that you need to provide an example of where you have already demonstrated the skills that are required under the job description or person specification in a previous role or situation. Do not fall into the trap of providing a 'generic' response that details what you 'would do' if the situation arose. Try to structure your responses in a logical and concise manner. The way to achieve this is to use the 'STAR' method of interview question response construction:

Situation

Start off your response to the interview question by explaining what the 'situation' was and who was involved.

Task

Once you have detailed the situation, explain what the 'task' was, or what needed to be done.

Action

Now explain what 'action' you took, and what action others took. Also explain why you took this particular course of action.

Result

Finally explain what the outcome or result was following your actions and those of others. Try to demonstrate in your response that the result was positive because of the action you took.

Finally, explain to the panel what you would do differently if the same situation arose again. It is good to be reflective at the end of your responses. This demonstrates a level of maturity and it will also show the panel that you are willing to learn from every experience.

THE DIFFERENT TYPES OF INTERVIEW QUESTIONS

Basically there are two different types of interview questions that you could be asked. I will try to explain each of them and what they mean:

1. Generic questions about you and your knowledge of the TOC and the Train Driver role.

Generic questions can be in any format. There is no particular structure to this type of question but they are generally far easier to respond to. Examples of generic questions would include:

- Why do you want to become a Train Driver?

- What has attracted you to this TOC in particular?

- What have you learnt about the role of a Train Driver?

- Why should we choose you over the other applicants?

2. Role related questions.

This type of question is more common during the structured interview and includes questions that are based around the job description/person specification. Examples of role related questions include:

- Being able to work under pressure.

- Following rules or guidelines.

- Providing a high level of customer care and service.

- Working as part of a team to achieve a task.

- Communicating a message to a group of people.

- Working with people from different backgrounds.

- Dealing with difficult and aggressive people.

On the following pages I have provided you with a number of sample interview questions and responses to assist you in your preparation. Please remember that the responses provided are not to be copied under any circumstances. Use them as a basis for your preparation taking examples from your own individual experiences and knowledge.

SAMPLE INTERVIEW QUESTIONS AND RESPONSES

Question 1 – Why do you want to become a Train Driver?

This question is inevitable, so it is important that you ensure you have a suitable answer prepared. Many people will respond with a standard answer such as "It's something that I've always wanted do since I was young". Whilst this is OK you need to back it up with genuine reasons that relate to the TOC you are applying for and other important reasons such as working in a customer-focused environment and a desire to learn new skills.

This type of question may be posed in a number of different formats such as the following:

Q.   Why do you want to become a Train Driver with our Company?

Q.   What has attracted you to the role of Train Driver?

Now take a look at the following sample response which will help you to prepare for this type of question. Once you have read it, use the template on the next page to create your own response based upon your own experiences and knowledge.

SAMPLE RESPONSE

Why do you want to become a Train Driver?

"I have wanted to become a Train Driver for many years now and have been preparing for the role for a long time. I have been very careful about which TOC to apply for and I have been impressed with the way your company operates. It sets itself high standards in terms of customer service and the safety standards that are expected of its employees. Apart from the fact that driving trains is quite an exciting job, I also very much enjoy new and different challenges. I understand that as a Train Driver there are a lot of new skills to learn, especially during the early years. The type of person I am means that I would work hard to ensure that I passed every exam first time. I also enjoy working in a customer-focused environment where a high level of service is essential. As a Train Driver you are responsible for the customer's safety and I would enjoy the high level of responsibility that comes with the position."

Template for Question 1 – Why do you want to become a train driver?

Question 2 – Why do you want to work for our company?

Once again this is a question that is likely to come up during your interview. In order to prepare for this question you need to carry out some research about the TOC you are applying for. The best place to get this information is via their website. See the Useful Contacts section for a list of current TOCs.

When responding to this type of question, try to focus on the positive aspects of the company's work. Do they run any customer-focused initiatives or have they won any awards for quality of work or service? It is always good to make reference to the positive aspects of their work, but do not mention any current or previous bad press. On the following page I have provided a sample response to this question to help you prepare. I have used Southern Rail as an example when constructing the response. Once you have read it, take the time to construct your own answer using the template provided.

SAMPLE RESPONSE

Why do you want to work for our company?

"I have been looking at a number of TOCs and I have been especially impressed with Southern Rail. The 'Meet the Managers' programme gives passengers the chance to meet Senior Managers and Directors and talk with them about the service. This demonstrates a high level of customer focus and care and I want to work for such a company as I believe I can bring the same high standards to the team.

I also understand that, over the next two years, Southern Rail aims to create a company that not only looks and feels different, but provides passengers with a better travelling experience. I believe that, whilst working with Southern Rail, I would have excellent career opportunities and therefore be very happy in my role as a Train Driver."

Template for Question 2 – Why do you want to work for our company?

Question 3 – What can you tell us about the role of a Train Driver?

You must be well prepared for this question prior to your interview. If you don't know what the role involves, then you shouldn't be applying for the post. When responding to this question, make sure you make reference to the job/person specification for the role. The job specification is a 'blueprint' for the role that you will be required to perform whilst working as a Train Driver. Therefore, it is essential that you know it. An example of a Train Driver's duties/ person specification is detailed below:

Person Specification

We need people who will share our passion, enthusiasm and commitment to deliver the best train service for our passengers. You will have outstanding communication skills and be strong on customer service. You must be able to learn and follow rigorous railway and safety procedures. Aged between 21-49, you will need to be decisive and have excellent concentration skills.

Job Description

Drive trains taking particular account of all permissible, temporary and emergency speed restrictions. Trains must be driven with regard to punctuality and customer comfort. Obey all fixed signals and hand signals. Communicate effectively with signallers and hand signallers regarding the transmission of verbal messages. The use of phonetic alphabet is mandatory. Ensure that customers are advised, either directly or through others, regarding train running matters. Ensure that your traction and route cards are kept up to date.

Now take a look at the sample response on the following page before constructing your own response using the template provided.

SAMPLE RESPONSE

What can you tell us about the role of a Train Driver?

"I understand that the role involves a high level of responsibility, concentration and lone working. To begin with, Train Drivers are responsible for ensuring that they drive trains taking particular account of all permissible, temporary and emergency speed restrictions.

The trains must be driven on time and drivers must ensure that they obey all fixed signals and hand signals as the safety of the trains and passengers is paramount. Part of the role also involves communicating with signallers and hand signallers regarding the transmission of verbal messages. Other elements of communication involve ensuring that customers are advised, either directly or through others, regarding train running matters.

Finally, Train Drivers must ensure that their traction and route cards are kept up to date. Safety is essential to the role of a competent Train Driver and keeping up to date with procedures and regulations is very important."

Template for Question 3 – What can you tell us about the role of a train driver?

Question 4 – What skills do you possess that you think would be an asset to our team?

When responding to questions of this nature, try to match your skills with the skills that are required of a Train Driver. On some TOC websites, you will be able to see the type of person they are looking to employ, usually in the recruitment section.

An example of this would be: 'We are looking for friendly, supportive people who share our professional, customer-focused approach. You must be a good team player with a flexible attitude and a willingness to learn.' Just by looking at the TOC's website, you should be able to obtain some clues as to the type of person they are seeking to employ. Try to think of the skills that are required to perform the role you are applying for and include them in your response.

The following is a sample response to the question. Once you have read it, take the time to construct your own response using the template provided.

SAMPLE RESPONSE
What skills do you possess that you think would be an asset to our team?

"I am a very conscientious person who takes the time to learn and develop new skills correctly. I have vast experience working in a customer-focused environment and fully understand that customer satisfaction is important. Without the customer there would be no company, so it is important that every member of the team works towards providing a high level of service.

I believe I have the skills, knowledge and experience to do this. I am a very good team player and can always be relied upon to carry out my role to the highest of standards. I am a flexible person and understand that there is a need to be available at short notice to cover duties if required. In addition to these skills and attributes, I am a very good communicator. I have experience of having to communicate to customers in my previous role and believe that this would be an asset in the role of a Train Driver. I am highly safety conscious and have a health and safety qualification to my name. Therefore, I can be relied upon to perform all procedures relevant to the codes of conduct and will not put myself or others in any danger whatsoever. Finally, I am very good at learning new skills which means that I will work hard to pass all of my exams if I am successful in becoming a trainee Train Driver."

Template for Question 4 – What skills do you possess that you think would be an asset to our team?

Question 5 – Can you tell us about a situation when you have had to work under pressure?

The role of a Train Driver will sometimes involve a requirement to work under pressure. Therefore, the recruitment staff want to know that you have the ability to perform in such an environment. If you have experience of working under pressure then you are far more likely to succeed as a Train Driver. When responding to a question of this nature, try to provide an actual example of where you have achieved a task whilst being under pressure. Questions of this nature are sometimes included in the application form, so try to use a different example for the interview.

I have provided you with a sample response to this question. Once you have read it, take the time to construct your own response based on your own individual experiences and knowledge using the template provided.

SAMPLE RESPONSE

Can you tell us about a situation when you have had to work under pressure?

"Yes, I can. In my current job as a car mechanic for a well-known company, I was presented with a difficult and pressurised situation. A member of the team had made a mistake and had fitted a number of wrong components to a car. The car in question was due to be picked up at 2pm and the customer had stated how important it was that his car was ready on time because he had an important meeting to attend.

We only had two hours in which to resolve the issue and I volunteered to be the one who would carry out the work on the car. The problem was that we had three other customers in the workshop waiting for their cars too, so I was the only person who could be spared at that particular time. I worked solidly for the next two hours making sure that I meticulously carried out each task in line with our operating procedures. Even though I didn't finish the car until 2.10pm, I managed to achieve a very difficult task under pressurised conditions whilst keeping strictly to procedures and regulations."

Template for Question 5 – Can you tell us about a situation when you had to work under pressure?

Question 6 – Can you tell me about a time when you have worked as part of a team to achieve a goal?

Having the ability to work as part of a team is very important to the role of a Train Driver. Train Operating Companies employ many people in different roles from Conductors to platform staff and from ticket office staff to caterers. In fact it is not uncommon for thousands of people to work for one particular TOC. Therefore, it is essential that every member of the team works together in order to achieve the ultimate goal of providing a high quality rail service. The recruitment staff will want to be certain that you can work effectively as part of a team, which is why you may be asked questions that relate to your team working experience.

There now follows a sample response to this question. Once you have read it, take time to construct your own response using the template provided.

SAMPLE RESPONSE

Can you tell me about a time when you have worked as part of a team to achieve a goal?

"Yes, I can. I like to keep fit and healthy and as part of this aim I play football for a local Sunday team. We had worked very hard to get to the cup final and we were faced with playing a very good opposition team who had recently won the league title. After only ten minutes of play, one of our players was sent off and we conceded a penalty as a result. Being one goal down and 80 minutes left to play we were faced with a mountain to climb. However, we all remembered our training and worked very hard in order to prevent any more goals being scored. Due to playing with ten players, I had to switch positions and play as a defender, something that I am not used to. The team worked brilliantly to hold off any further opposing goals and after 60 minutes we managed to get an equaliser. The game went to penalties in the end and we managed to win the cup. I believe I am an excellent team player and can always be relied upon to work as an effective team member at all times. I understand that being an effective team member is very important if the Train Operating Company is to provide a high level of service to the passenger. However, above all of this, effective teamwork is essential in order to maintain the high safety standards that are set."

Template for Question 6 – Can you tell me about a time when you have worked as part of a team to achieve a goal?

Question 7 – Can you provide us with an example of a project you have had to complete and the obstacles you had to overcome?

Having the ability to complete tasks and projects successfully demonstrates that you have the ability to complete your trainee Train Driving course. Many people give up on things in life and fail to achieve their goals. The recruitment staff need to be convinced that you are going to complete all training successfully and, if you can provide evidence of where you have already done this, then this will go in your favour.

When responding to this type of question, try to think of a difficult, drawn out task that you achieved despite a number of obstacles that were in your way. You may choose to use examples from your work life or even from some recent academic work that you have carried out. Take a look at the following sample question before using the template provided to construct your own response based on your own experiences.

SAMPLE RESPONSE

Can you provide us with an example of a project you have had to complete and the obstacles you had to overcome?

"Yes I can. I recently successfully completed a NEBOSH course (National Examination Board in Occupational Safety and Health) via distance learning. The course took two years to complete in total and I had to carry out all studying in my own time whilst holding down my current job.

The biggest obstacle I had to overcome was finding the time to complete the work to the high standard that I wanted to achieve. I decided to manage my time effectively and I allocated two hours every evening of the working week in which to complete the work required. I found the time management difficult but I stuck with it and I was determined to complete the course. In the end I achieved very good results and I very much enjoyed the experience and challenge. I have a determined nature and I have the ability to concentrate for long periods of time when required. I can be relied upon to finish projects to a high standard."

Template for Question 7 – Can you provide us with an example of a project you had to complete and the obstacle you had to overcome?

Question 8 – Can you provide us with an example of a safety-related task that you have had to perform?

Safety is an extremely important part of the Train Driver's role, and the recruitment staff need to know that you are capable of working safely at all times. The term 'safety' should be an integral part of your responses during the interview. Making reference to the fact that you are aware of the importance of safety at every opportunity is a positive thing. When responding to safety-related questions try to include examples where you have had to work to, or follow, safety guidelines or procedures. If you have a safety qualification then it is definitely worthwhile mentioning this during your interview. Any relevant safety experience or related role should also be discussed.

Now take a look at the following sample response before using the template provided to construct your own response.

SAMPLE RESPONSE

Can you provide us with an example of a safety-related task that you have had to perform?

"I currently work as a gas fitter and I am often required to perform safety-related tasks. An example of one of these tasks would involve the installation of gas-fired boilers. When fitting a gas boiler I have to ensure that I carry out a number of safety checks during the installation stage which ensures my work is safe and to a high standard. I have been trained, and I am qualified, to carry out my work in accordance with strict safety guidelines. I also have a number of safety certificates to demonstrate my competence.

I am fully aware that if I do not carry out my job in accordance with safety guidelines there is the possibility that somebody may become injured or even killed."

Template for Question 8 – Can you provide us with an example of a safety related task that you have had to perform?

Hopefully you are now starting to get a feel for how you need to respond to the interview questions. The following set of interview questions are further examples of questions you could get asked during the interviews.

Question 9 – How do you think you would cope with working on your own for long periods?

Lone working is an unfortunate part of the Train Driver's job. You will spend many hours on your own and this can be a problem for many people. You need to think carefully about this downside to the job. Can you cope with it? Do you have any experience of working on your own? If you do not then you will have to convince the panel that you can cope with it.

Question 10 – What is your sickness record like and what do you think is an acceptable level of sickness?

Most employers detest sickness and they especially detest sickness that is not genuine. Every day that an employee is off sick will cost the TOC dearly. Therefore, they want to employ people who have a good sickness record. Obviously you cannot lie when responding to this question as the TOC will carry out checks. The latter part of the question is simple to answer. Basically no amount of sickness is acceptable but sometimes genuine sickness cannot be helped. Remember to tell them that you do not take time off sick unless absolutely necessary and you can be relied upon to come to work.

Question 11 – Have you ever worked during the night and how do you feel about working shifts?

Train Driving involves irregular shifts and the Train Operating Company want to know that you can handle them. Speak to any person who works shifts and they will tell you that after a number of years they can start to take their toll. Remember to tell the panel that you are looking forward to working shifts and in particular night duties. If you can provide examples of where you have worked irregular shift patterns then remember to tell them.

Question 12 – Would you get bored of driving the same route day in, day out?

Of course the only answer here is no! Yes, we would all probably get bored of the same journey everyday, but don't tell them this.

Question 13 – How many people work for this TOC?

Questions that relate to facts and figures about the TOC might come up. They want to know that you are serious about joining them and that you are not just there to become a Train Driver. Make sure you study their website and find out as much about them as possible.

Question 14 – How many stations does the company service?

Once again, this is a question that relates to your knowledge of the TOC. This kind of information can usually be found by visiting their website. Please see our Useful Contacts section for more details.

Question 15 – What are the mission and aims of this company?

Many organisations including Train Operating Companies set themselves aims and objectives. They usually relate to the high level of customer service that they promise to deliver. When you apply to become a Train Driver you should not only prepare for each stage of the selection process but you should also learn as much as possible about the company you are applying to join. Learning this kind of information is important and it will demonstrate your seriousness about joining their particular company. Always remember this rule, working for the TOC comes first, becoming a Train Driver comes second! Visit the website of the TOC in order to view their mission, aims, objectives or customer charter.

Question 16 – Can you provide us with an example of when you have had to work in an emergency?

This question is also likely to be asked during the application form stage of the process. Being able to remain calm under pressure is very important and will form an integral part of your training. Maybe you have had to deal with an emergency at work or even in the home? Whatever example you decide to use, make sure you tell them that you stayed calm and focused on the task in hand. Make reference to the importance of safety during your response too.

Question 17 – Do you think it's important for staff to wear a uniform?

The answer to this question should be yes. The reason for this is that a uniform gives customers and passenger's confidence in the service they are receiving. It is also important during an emergency situation so that customers know who to turn to for help and guidance. Uniforms are positive for the image of Train Operating Companies which is why they use them. Be positive about uniforms and tell them that you are looking forward to wearing one and taking pride in your appearance.

Question 18 – If you were a train driver and you came across an obstacle on the track, what would you do and why?

This type of question is rare but it has been asked during the selection interview. Basically, you need to ensure that you tell the panel that you would follow the procedures you learnt during your training. You should tell them that the safety of your passengers and the train would be paramount. You would raise the alarm immediately by contacting the control centre so that other trains could also be informed of the danger and they could then take appropriate action as necessary. During your responses you should never compromise safety!

FINAL TIPS FOR PREPARING FOR THE INTERVIEWS

- Make sure you turn up to your interview on time! Find out the route to the interview location well in advance and make sure you don't get stuck in traffic or have any problems parking. Prepare for these eventualities well in advance.

- Wear formal clothing for your interview. Make sure you are clean-shaven and your shoes are clean and polished. Remember that you will be representing the company if you are successful and your appearance is very important.

- Visit the website of the TOC you are applying for and learn information about how they operate and what they are about. This is important so that you can create an image that you are serious about working for them and not just interested in becoming a Train Driver.

- During your preparation for the interview, try to think of some recent examples of situations you have been in that are relevant to the role of a Train Driver.

- When responding to the questions try to concentrate on what you have achieved so far during your life. It is important that you can demonstrate a track record of achievement.

- Make sure you smile during your interview. Sit up straight in the chair and do not fidget.

USEFUL CONTACTS

Within this section of the guide I have provided you with a list of Train Operating Companies that exist in England, Scotland and Wales. Please note, the list is not exhaustive and you may find other TOCs operating within your area/region. Some of the contact details may also change from time to time.

Arriva Trains Wales
www.arrivatrainswales.co.uk
Arriva Trains Wales
St Mary's House
47 Penarth Road
Cardiff CF10 5DJ
0845 6061 660
customer.relations@arrivatrainswales.co.uk

C2C
www.c2c-online.co.uk
10th Floor, 207 Old Street
London EC1V 9NR
0845 6014873
c2c.customerrelations@nationalexpress.com

Chiltern Railways
www.chilternrailways.co.uk
Chiltern Railways 2nd floor,
Western House
Rickfords Hill
Aylesbury
Buckinghamshire HP20 2RX
08456 005 165

CrossCountry
www.crosscountrytrains.co.uk
CrossCountry
Cannon House
18 The Priory
Queensway
Birmingham B4 6BS
0870 010 0084
info@crosscountrytrains.co.uk

East Coast
www.eastcoast.co.uk
East Coast House
25 Skeldergate House
York Y01 6DH
08457 225 225

East Midlands Trains
www.eastmidlandstrains.co.uk
1 Prospect Place
Millennium Way
Pride Park
Derby DE24 8HG
08457 125 678
getintouch@eastmidlandstrains.co.uk

Eurostar
www.eurostar.com
Times House
Bravingtons Walk
Regent Quarter
London N1 9AW
08701 606 600

First Great Western

www.firstgreatwestern.co.uk
Head Office
Milford House
1 Milford Street
Swindon SN1 1HL
08457 000 125

First Hull Trains

www.hulltrains.co.uk
First Hull Trains
FREEPOST
RLYY-XSTG-YXCK
4th Floor
Europa House
184 Ferensway
Hull HU1 3UT
08456 769 905

First TransPennine Express

www.tpexpress.co.uk
Floor 7
Bridgewater House
60 Whitworth Street
Manchester M1 6LT
0845 600 1671

Gatwick Express

www.gatwickexpress.com
Go Ahead House
26-28 Addiscombe Road
Croydon CR9 5GA
0845 850 15 30

Grand Central
www.grandcentralrail.co.uk
River House
17 Museum Street
York YO1 7DJ
0845 6034852
info@grandcentralrail.com

Heathrow Connect
www.heathrowconnect.com

Heathrow Express
www.heathrowexpress.com
6th Floor
50 Eastbourne Terrace
Paddington
London W2 6LX
020 8750 6600

Island Line Trains
www.southwesttrains.co.uk
Ryde St Johns Road Stations
Ryde, Isle of Wight PO33 2BA
01983 812 591

London Midland
www.londonmidland.com
102 New Street
Birmingham B2 4JB
0121 634 2040
comments@londonmidland.com

London Overground
www.tfl.gov.uk/overground
London Overground Rail Operations
125 Finchley Road
London NW3 6HY
0845 601 4867

London Underground
www.tfl.gov.uk

Merseyrail
www.merseyrail.org
Rail House
Lord Nelson Street
Liverpool L1 1JF
0151 702 2534

National Express East Anglia
www.nationalexpresseastanglia.com
Floor One
Oliver's Yard
55 City Road
London EC1Y 1HQ
0845 600 7245
nxea.customerrelations@nationalexpress.com

Northern Rail
www.northernrail.org
Northern Rail Ltd
Northern House
9 Rougier Street
York YO1 6HZ
0845 00 00 125
customer.relations@northernrail.org

ScotRail
www.scotrail.co.uk
Atrium Court
50 Waterloo Street
Glasgow G2 6HQ
08700 005151

South West Trains
www.southwesttrains.co.uk
South West Trains
Friars Bridge Court
41-45 Blackfriars Road
London SE1 8NZ
08700 00 5151

Southeastern
www.southeasternrailway.co.uk
Customer Services
PO Box 63428
London SE1P 5FD
0845 000 2222

Stansted Express
www.stanstedexpress.com

Virgin Trains
www.virgintrains.co.uk
Virgin Trains
85 Smallbrook
Queensway
Birmingham B5 4HA
0845 000 8000

OTHER USEFUL WEBSITES

How to Become a Train Driver
The following website is the UK's leading Train Driver information website. It is an invaluable source of information and one that I strongly recommend you visit and study:

www.traindriver.org

Rail Safety & Standards Board
www.rssb.co.uk

The Department for Transport
www.dft.gov.uk

Rail Technical Pages
www.railway-technical.com

Track Access – Route Learning website
www.trackaccess.co.uk

FOR MORE CAREERS GUIDANCE GO TO:

WWW.HOW2BECOME.COM

Get Access To

FREE

TRAIN DRIVER

TEST QUESTIONS

www.MyPsychometricTests.co.uk

Printed and bound by CPI Group (UK) Ltd, Croydon, CR0 4YY

05/03/2026

02065671-0001